Charles Adiel Lewis Totten

Joshua's Long Day and the Dial of Ahaz

A scientific vindication

Charles Adiel Lewis Totten

Joshua's Long Day and the Dial of Ahaz
A scientific vindication

ISBN/EAN: 9783337417260

Printed in Europe, USA, Canada, Australia, Japan

Cover: Foto ©berggeist007 / pixelio.de

More available books at **www.hansebooks.com**

JOSHUA'S
LONG DAY

AND

THE DIAL OF AHAZ.

A SCIENTIFIC VINDICATION

AND

"A MIDNIGHT CRY."

BY

CHARLES A. L. TOTTEN,

FIRST LIEUTENANT FOURTH ARTILLERY, U. S. A.; PROFESSOR OF MILITARY
SCIENCE AND TACTICS, S. S. S. OF YALE UNIVERSITY;
AUTHOR OF "STRATEGOS;" "AN IMPORTANT QUESTION;" "FACTS,
FANCIES, LEGENDS, AND LORE OF NATIVITY;" ETC.
EDITOR OF "OUR RACE," ETC.

THE PARABLE OF THE FIG TREE:
"When her branch is yet tender, and putteth forth leaves, ye know that Summer is near: So ye in like manner, when ye shall see these things come to pass, know that it is nigh, even at the doors." Mark xiii. 28-29.

NEW HAVEN, CONN.:
THE OUR RACE PUBLISHING COMPANY.
1890.

COPYRIGHTED, 1890,
BY
CHARLES A. L. TOTTEN.
(*All rights reserved.*)

A WORD TO "THE WISE."

Dan. xii. 10; Matt. xxv. 1-13.

" The Sun and the Moon stood still in their habitation."   *Hab.* iii. 11.

STUDY No. 2

OF

THE OUR RACE SERIES.

THE VOICE OF HISTORY.

TOTTEN.

"And God said Let there be lights in the firmament of the heaven to divide the day from the night; and let them be for signs and for seasons, and for days, and years:

And let them be for lights in the firmament of the heaven to give light upon the earth, and it was so. And God made two great lights; the greater light to rule the day and the lesser light to rule the night."   Gen. i., 14-15.

STUDY NUMBER TWO.

THE VOICE OF HISTORY.

JOSHUA'S LONG DAY, AND THE DIAL OF AHAZ.

A SCIENTIFIC VINDICATION, AND A MIDNIGHT CRY.

CONTENTS.

	PAGE
PREFACE,	xi
INTRODUCTION,	xiii
THE BIBLICAL RECORD,	1
JOSHUA'S LONG DAY,	5
THE SHADOW ON THE DIAL,	11
THE ELEMENTS AS VERIFIED,	17

PART I.—DISCUSSION.

	PAGE		PAGE
THE ARGUMENT,	23	PROVED FROM THE ALMANAC,	67
SECULAR CORROBORATION,	25	THE APOLOGISTS ARRAIGNED,	71
INCEPTION OF PROBLEM,	27	THE SWORD OF DAMOCLES,	72
REFERENCE TO MOON ANOMALOUS,	31	THE TRUE CHRONOLOGY,	74
REAL EFFECT OF STOPPAGE,	33	CREATION'S 6,000 YEARS,	75
SOLI-LUNAR CYCLES,	36	JOSHUA, CHRIST, COLUMBUS,	76
TRUE ORIGIN OF "TIME,"	37	THE KEY TO CHRONOLOGY,	77
THE INTERCALATED DAY,	38	THE WEEK UNBROKEN,	78
THE SITE UNIQUE,	40	THE END OF THE AGE,	80
THE BATTLE DESCRIBED,	41	A SIGNIFICANT YEAR,	82
WITHOUT SUNSET, NO SUNRISE,	44	A SOLEMN WARNING,	83
A MILITARY COMMENTARY,	47	JEWISH IRREDENTALISM,	85
THE CONQUEST OF PALESTINE,	51	THE LAST KING OF THE FRANKS,	85
CORROBORATED BY ECLIPSES,	53		
VERIFIED BY EQUINOXES,	59	THE CONTROVERSY OF ZION,	89
SHADOWED ON THE DIAL,	60	A MIDNIGHT CRY,	90

CONTENTS.

PART II.—APPENDICES.

	PAGE		PAGE
A. THE BOOK OF JASHER,	95	*F*. CHRONOLOGICAL ERAS HARMONIZED,	110
B. CASUAL ECLIPSES,	97		
C. EARLIEST AND LATEST ECLIPSE,	97	*G*. ASTRONOMICAL FORECAST (1891-2)	115
D. TIMES AND SEASONS CHANGED,	101	*H*. CAIAPHAS AND LEO XIII,	117
E. BIBLICAL CYCLES ARE ALL ASTRONOMICAL,	105	*I*. JUDAH HOMEWARD BOUND,	120

PART III.—CHRONOLOGICAL APPENDICES.

	PAGE
THE CHALDEE BABYLONIAN ERA, INTRODUCTION,	141
PRELIMINARY CHRONOLOGICAL OUTLINE,	143
ORIGIN OF "TIMES OF THE GENTILES,"	155
THE CHRONOLOGY OF JEREMIAH VINDICATED,	175
PRELIMINARY OUTLINE,	179
CHRONOLOGICAL ARRANGEMENT,	187
THE END OF THE AGE,	201
THE CONCLUSION OF THE MATTER,	209

TABLES.

	PAGE
THE 176TH METONIC CYCLE,	69
A. M., A. D., B. C., A. U. C., ETC., YEARS SYNCHRONIZED,	113
CHRONOLOGY OF "II ASSYRIAN EMPIRE,"	141
HOSHEA'S REIGN HARMONIZED,	143
OPENING CHAPTER OF "TIMES OF THE GENTILES,"	155
CHRONOLOGY OF JEREMIAH, OPENING CHAPTER,	179
" " CLOSING CHAPTER,	187
THE BIRTH-DAY OF TIME,	205
THE END OF THE AGE,	207
GENERAL CHART OF THE 2,520 YEARS,	215

ILLUSTRATIONS.

THE BATTLE OF BETH HORON. MILITARY MAP,	FRONTISPIECE
ILLUSTRATION, DIAGRAM OF ASTRONOMICAL ELEMENTS,	page 16
EDITORIALS,	217
MISCELLANEOUS,	227

PREFACE.

Having dedicated this Volume in the spirit of Matthew xxv. 1-13, and of Daniel xii. 10, it is meet for me to preface it with equal brevity.

What, therefore, I have to say may be summarized as follows:

The Church has mourned long enough, for the World has not lamented; it is on this account that I have "piped" up a scientific tune in the Congregation, to see if "his terrestrial majesty" will warm up to the "dance" (Matt. xi. 16-17).

In the meantime let it be pointedly stated that if any one shall find the clear and simple logic and arithmetic of this volume too forbidding for even an effort at its comprehension, it is but a proof positive that he will also find the "Needle's Eye" too narrow to admit even his own dimensions (Mark x. 23-27).

In this study I have taken the two most doubted texts of Holy Writ as my theme, and, having shown that they agree with the calendar, square with the truth, and complement each other, I have endeavored to raise the alarm which suits the hour that is just striking upon the Dial of the Ages (Isa. xxi. 5, 11 ; lii. 8. Jer. vi. 17; xxxi. 6. Ezek. iii. 17; xxxiii. Hab. ii. 1-4).

The earnest response which has greeted the First Study of the Our Race Series, has fully demonstrated that God has left unto himself his thousands in Israel who have not bowed their knees to Baal, and many mouths that have not kissed him, nor lent themselves unto the delusions of faithless doctrines (I. Kings, xix. 18).

I thank them all for their numerous letters of assistance and encouragement, and submit to them that the hour has now come when we must spread our knowledge of Truth with one consent, and propagate the Midnight Cry! (Luke xii. 35-40).

C. A. L. T.

NEW HAVEN, CONN.:
    Harvest Moon, Sept., 1890, A. D.

INTRODUCTION.

"THERE never was a day in which earnest Bible study was more necessary than the present one, nor was there ever a day in which so rich a harvest stood ready for the reaping. Most religious people are literally afraid to investigate the Bible, and well they may be if the canons of the 'Higher Criticism' are to guide their study. Most of the laity consider it to be beyond their sphere, and so far as 'Moses and the Prophets' are concerned, even the clergy almost entirely neglect them.

"We readily grant that Sin, Repentance, and the Gospel of a *Saviour* are the vital 'ends' of apostolic work. Nevertheless we hold that Christ and the Resurrection *cannot* be successfully preached *in this age* upon the undermined foundation left by the Higher Critics. It is well for them if they can hold their own souls within the fold; we question it; but be this as it may, it is the *rest of men* that are the ones whom Christ desires to save, and *they* have logic left, and cannot be savingly reached by any other means than a logical exegesis of the whole Bible, and a satisfactory explanation of its inspiration as such, upon the basis that it is '*the truth, the whole truth, and nothing but the truth.*' For, not though one arose from the dead, will men *believe*, unless they likewise are taught to *believe* implicitly, and are made to *understand*, 'Moses and the Prophets.'

" It is the Bible that Atheists and Infidels attack, —the Old Testament chiefly,—for they are logical, and perceive that if the foundation goes, the superstructure cannot stand, no matter how eloquently it can be clothed in Agnostic sermons. Hence this Old Testament is our one and only bulwark of defense, and the Romance of History will make of him who reads 'Moses and the Prophets' in the light of Anglo-Saxon facts. a GNOSTIC indeed, and one who can fully show whereon he stands, and *why* he 'knows.'

" It will not do to preach Christ and deny Moses. It will not do to doubt the universality of the Flood, and ask men to accept a Saviour who alludes to it! It will not do to doubt Joshua's Long Day, with the sun and moon poised in midheaven while he fought, and yet stultify our hearts with hopes of a LONGER DAY when even sun and moon will not be needed ! If the story of Eden and the Deluge, of Jericho and Joshua, are myths, or fables, and not literal facts, then, to the still *rational* mind, all that follows them is equally so, and faith, lost in those who foretold his advent, can never be savingly and logically found again in Christ and his apostles.

"If, therefore, we are to resume our place

INTRODUCTION. xv

militantly among the noble army of those who have already testified for Jesus Christ with their lives and works, we must repudiate *in toto* this iniquitous school of criticism, and recapture, somehow or other, the Ararat redoubt, replace the Long Day in our scientific chronology, believe Moses rather than the Moabite stone, and the Bible rather than a sunburned brick dug up at Babylon.

"As the study of prophecy was impressively recommended by the Saviour, we must *study* it, and do so until we understand it; but in no wise may we dare to alter it in jot or tittle!"

We extract the foregoing from Study No. 1, of " Our Race, Its Origin and Destiny," as a fitting introduction to the present volume—Study No. 2 —which we send out to supplement the work begun in the former.

The times in which we live are rushing too rapidly to their culmination to permit of adding leaves to leaves, and smothering simple facts beneath the winter garments of verbosity. We, therefore, make no apology for the working clothes in which these notes appear, and are too anxious to see them in the hands of " Our Race" to delay them longer in our own.

If those to whom this rectified chronology shall come " can receive it," it has a momentous significance upon the " prophetic times " which yet remain,—and whose abrupt ending is now apparently so imminent.

That, after its perusal, some at least—"the wise"—may better understand the import of "the half hour of silence" (Rev. viii. 1) which has lately so surprised "the watchers," the author is convinced, and he will be satisfied if it shall *"witness"* to the rest, the certainty of what will follow soon.

Yet, let it not be understood that the author pretends to be among the prophets (1 Sam. x. 11-12, xix. 24). He has no such aspirations. But he does claim all the rights and authority which pertain to all who live in these latter generations, whose duty it is to study Moses and the prophets, and is determined to give the benefit of some of his researches to others engaged in the same absorbing occupation, and so, as it were, to exchange notes with what, it is to be hoped, is a growing number of Godly men who love the same pursuit.

The only foundation upon which to understand either History or Prophecy is a correct Calendar of the "Times and the Seasons," or rather *the* correct Calendar, for, from the very nature of *Time* there can be but one sequence to it, and that will be a sequence through which all the cycles of heaven will reverse accurately. Upon such a system we can fairly hope to work, but upon no other. And upon it we must know not only the dates *at* which the prophecies of Scripture were uttered, but the dates *from* which, and *to* which, *they* themselves refer, in order that we may in any sort of confidence even attempt to interpret the "sure word of prophecy" aright.

The trouble heretofore has been that we have been in doubt upon each of these points, and so the valuable labors of nearly all the College of Historico-Prophetical Students have been mere "approximations" after all.

And thus "the Church," which has waited on their efforts, has grown weary of the topic, and has almost closed the canon in despair.

Hence, in re-opening the subject, we shall do so *ab ovo*, and shall first endeavor to show that certain essentials hitherto neglected must be faithfully put back into Israel's chronology in order to rectify IT, and thereafter we shall call attention to certain inevitable deductions which seem to be at least portentious enough to warrant a renewed appeal unto all concerned.

Yet, be this as it may, or rather let it strike others as it will, it is none the less the duty of one who has come honestly by such convictions as control the present writer, and can show cause for his deductions, to give them all the publicity he can, for certainly his own conscience would not hold him guiltless did he conceal such knowledge from his imperilled fellows—he so believing them to be—and his condemnation could but be commensurate with the magnitude of the peril as he saw it.

The desire to rectify the accepted Calendar of Our Race's history, by reinstating therein its two most doubted events, has resulted not only in a most interesting return to the mere Chronologist as such, but has enabled us to re-write two mutually corroborating chapters which commence together at the most important "Era" of human affairs, and run parallel to each other long enough to demonstrate their accuracy.

But this "Era," from which they both start, is the all important *origine* of the "Times of the Gentiles," and if its beginning is known its latter end may of course be calculated. Now, it is the result of this calculation which is so particularly portentious, for—according to rules which have had the unanimous consent of all who have ever devoted themselves to this deep subject,—the "end of the present dispensation" terminates with the century now waning, and "the beginning of the end" dates officially with the autumnal Equinox (1890 A. D.) with which the issue of this present volume "happens" to synchronize!

We were not present when the "Times and Seasons" were instituted, but One was to whom their "speech" is still without confusion, and it is enough for us, who live now in "these latter days," to pray that we may stand in our own "lot" when they terminate. Whenever that may be, it agrees with faith, and with human charity, to recognize that it will have compassed the last moment of "Grace" which the Divine mind shall deem necessary, and in the meanwhile it behooves all *wise men* not only to be wide-awake, but also to " go forth to meet Him "—and to *tarry there, henceforth, until He comes!*

It is the consensus of the *truly* "Christian world" that not a "sign" but ONE (2 Thess. ii.) is now wanting, and the rest of the world is quite enough exercised, about they know not what, to demonstrate that "Spring" is very close at hand.

"*And it shall come to pass in that day, saith the Lord God, that I will cause the sun to go down at noon, and I will darken the earth in the clear day.*" Amos viii. 9.

THE BIBLICAL RECORDS,

AND

THE ELEMENTS OF THEIR MODERN VERIFICATION.

" *Who commandeth the sun and it riseth not;
and sealeth up the stars.*" *Job* ix. 7.

THE BIBLICAL ACCOUNT

OF

JOSHUA'S LONG DAY.

"The Battle of Beth Horon," remarks Dean Stanley, "is one of the most important in the history of the world; and yet the very name of this great battle is far less known to most of us than that of Marathon or Cannae." (*Dict. of Bib.* art. Beth Horon; Stanley, p. 208).

"Beth Horon (the *House of Caverns*) was the name of two villages, an upper, and a "nether" or lower (Josh. xvi. 3-5; Chron. vii. 24), on the steep road from Gibeon to Azekah, and the Philistine Plain, (Josh. x. 10-11, 1 Macc. iii. 24), which is still the great road of communication from the interior of the country to the sea-coast. The two Beth Horons still survive in the modern villages of *Beit-ur, el* tahta and *el* foha. On this same spot Judas Maccabæus won a great victory over the forces of Syria under Seron (1 Macc. iii. 13-24), and still later the Roman army under Cestius Gallus was totally cut up (Josephus, B. 11, 19 § 8-9)." Smith's Old Test. Hist. Dr. Smith does not seem to credit the chief event of this battle, since he says "The miracle must be understood as *phenomenal*, namely, that the sun and moon *appeared* to the Israelites to stand still" (!?) Now, this is simply begging the question, and begging with it every other miracle mentioned in the Bible! Most commentators regard the matter as a mere quotation from a poetical book called Jasher,* and, without exception, so far as the author knows or can find out, the Theological library of to-day contains no volume in which the absolute integrity of the account is candidly admitted and fairly argued.

The result is that this battle, so decisive from a military stand-point, and marked by so stupendous a FACT from the historical, chronological and astronomical points of view, has fallen entirely out of serious thought, and now-a-days serves merely as a text wherewith to point the shaft of ridicule and doubt. Indeed, upon the basis of the generally accepted "poetical version" of this incident, we much prefer the out-and-out position of Renan himself, as given in his "History of Israel" (q. v.), and with it, were we honestly convinced of the reliability of that method of treating the Scriptures,—we would logically go to the full extreme and reject its Divine inspiration *in toto*. But the wise man cannot do this; and so, to conserve his reason, he is forced to go down into the depths of all things

---
* *Vide* Appendix A.

until the truth shines out—convinced that it is there, if but with patience it be sought.

It is on this account that we deem it necessary to preface our study of the subject in hand by quoting at length the Biblical records of the only two *alterations* of "Time" which pretend to have had the authority of Him who instituted both it and the mechanism which records it. We shall then see what it is that our arguments and demonstrations have to deal with—an interpolated 24 hours—and so, with the theorem fairly understood, we shall be fully equipped for our task.

JOSHUA X.

THE SUN AND MOON STAND STILL.

CHAPTER X.   *References.*

1 *Five kings war against Gibeon.* 6 *Joshua rescueth it.*
10 *God fighteth against them with hailstones.* 12 *The*   2554 A. M.
sun and moon stand still at the word of Joshua. 16   1442 B. C.
The five kings are mured in a cave. 23 *They are
brought forth,* 24 *scornfully used,* 26 *and hanged.*
28 *Seven kings more are conquered.* 43 *Joshua returneth to Gilgal.*

NOW it came to pass, when Adoni-zedec king of Jerusalem, had heard how Joshua had taken Ai, and had utterly destroyed it; [a] as he had done to   a Ch. 6. 21.
Jericho and her king, so he had done to [b] Ai and her   b Ch. 8. 22, 26, 28.
king; and [c] how the inhabitants of Gibeon had   c Ch. 9. 15.
made peace with Israel, and were among them:

THE VOICE OF HISTORY.

^d Exod. 15. 14,
15, 16.
Deut. 11. 25.
4 Heb. *cities of the kingdom.*

2 That they ^d feared greatly, because Gibeon *was* a great city, as one of the ⁴ royal cities, and because it *was* greater than Ai, and all the men thereof *were* mighty.

3 Wherefore Adoni-zedec king of Jerusalem sent unto Hoham king of Hebron, and unto Piram king of Jarmuth, and unto Japhia king of Lachish, and unto Debir king of Eglon, saying,

4 Come up unto me, and help me, that we may smite Gibeon: ^e for it hath made peace with Joshua and with the children of Israel.

e Ver. 1.
1. 9. 15.

2555 A. M.
1441 B. C

5 Therefore the five kings of the Amorites, the king of Jerusalem, the king of Hebron, the king of Jarmuth, the king of Lachish, the king of Eglon, ^f gathered themselves together, and went up, they and all their hosts, and encamped before Gibeon, and made war against it.

f Ch. 9. 2.

g Ch. 5. 10.
& 9.6.
Monday.

6 ¶ And the men of Gibeon sent unto Joshua ^g to the camp to Gilgal, saying, Slack not thy hand from thy servants; come up to us quickly, and save us, and help us: for all the kings of the Amorites that dwell in the mountains are gathered together against us.

h Ch. 8. 1.

7 So Joshua ascended from Gilgal, he, and ^h all the people of war with him, and all the mighty men of valor.

Ch. 11. 6.
Judg. 4. 14.

8 ¶ And the LORD said unto Joshua, ⁱ Fear them not: for I have delivered them into thine hand; ^k there shall not a man of them stand before thee.

k Ch. 1. 5.
Tuesday.

9 Joshua therefore came unto them suddenly, *and* went up from Gilgal all night.

l Judg. 4. 15.
1 Sam. 7. 10,
12.
Ps. 18. 14.
Isai. 28. 21.
m Ch. 16, 3, 5.
n Ch. 15. 35.

10 And the LORD ^l discomfited them before Israel, and slew them with a great slaughter at Gibeon, and chased them along the way that goeth up ^m to Beth-horon, and smote them to ⁿ Azekah, and unto Makkedah.

11 And it come to pass, as they fled from before Israel, *and* were in the going down to Beth-horon,

THE BIBLICAL ACCOUNT.

°that the LORD cast down great stones from heaven upon them unto Azekah, and they died: *they were* more which died with hailstones than *they* whom the children of Israel slew with the sword.

12 ¶ Then spake Joshua to the LORD in the day when the LORD delivered up the Amorites before the children of Israel, and he said in the sight of Israel, ᵖ Sun, ²stand thou still upon Gibeon; and thou, Moon, in the valley of ᵈ Ajalon.

13 And the sun stood still, and the moon stayed, until the people had avenged themselves upon their enemies. ʳ *Is* not this written in the book of ³Jasher? So the sun stood still in the midst of heaven, and hasted not to go down about a whole day.

14 And there was ˢ no day like that before it or after it, that the LORD hearkened unto the voice of a man: for ᵗ the LORD fought for Israel.

15 ¶ ᵘ And Joshua returned, and all Israel with him, unto the camp to Gilgal.

o Ps. 18. 13, 14. & 77. 17. Isai. 30. 30. Rev. 16. 21.

Tuesday.

p Isai. 23. 21. Hab. 3. 11.
2 Heb. *be silent.*
q Judg. 12. 12.

r 2 Sam. 1. 18.
3 Or, *the upright?*

Tues.—Wed.

s See Isai. 38. 8.

t ver. 42. Deut. 1. 30. Ch. 23. 3.
u ver. 43.

A RESUME,

DETAILING THE LEADING INCIDENTS OF THE LONG DAY.

16 But these five kings fled, and hid themselves in a cave at Makkedah.

17 And it was told Joshua, saying, The five kings are found hid in a cave at Makkedah.

18 And Joshua said, Roll great stones upon the mouth of the cave, and set men by it for to keep them:

19 And stay ye not, *but* pursue after your enemies, and ⁴ smite the hindmost of them; suffer them not to enter into their cities: for the LORD your God hath delivered them into your hand.

Tues.—Wed. 933,285–6 "Days of the World."

4 Heb. *cut off the tail.*

20 And it came to pass, when Joshua and the children of Israel had made an end of slaying them with a very great slaughter, till they were consumed, that the rest *which* remained of them entered into fenced cities.

21 And all the people returned to the camp to Joshua at Makkedah in peace: ˣ none moved his tongue against any of the children of Israel.

x Exod. 11. 7.

22 Then said Joshua, Open the mouth of the cave, and bring out those five kings unto me out of the cave.

23 And they did so, and brought forth those five kings unto him out of the cave, the king of Jerusalem, the king of Hebron, the king of Jarmuth, the king of Lachish, *and* the king of Eglon.

24 And it came to pass, when they brought out those kings unto Joshua, that Joshua called for all the men of Israel, and said unto the captains of the men of war which went with him, Come near, ʸ put your feet upon the necks of these kings. And they came near, and put their feet upon the necks of them.

y Ps. 107, 40. & 110. 5. & 149. 8. 9. Isai. 26. 5, 6. Mal. 4. 3.

25 And Joshua said unto them, ᶻ Fear not, nor be dismayed, be strong and of good courage: for ᵃ thus shall the LORD do to all your enemies against whom ye fight.

z Deut. 31. 6. 8. Ch. 1. 9.

2555 A. M. winter solstice 933,285-6. "Days of the World."

a Deut. 3. 21. & 6. 19.

26 And afterward Joshua smote them, and slew them, and hanged them on five trees: and they ᵇ were hanging upon the trees until the evening.

b Ch. 8. 29.

27 And it came to pass at the time of the going down of the sun, *that* Joshua commanded, and they ᶜ took them down off the trees, and cast them into the cave wherein they had been hid, and laid great stones in the cave's mouth. *which remain* until this very day.

933,287th day, Thursday.

c Deut. 21, 23. Ch. 8. 29.

Thursday.

28 ¶ And that day Joshua took Makkedah, and smote it with the edge of the sword, and the king thereof he utterly destroyed, them, and all the souls that *were* therein: he let none remain: and he did to the king of Makkedah ᵈ as he did unto the king of Jericho.

d Ch. 6. 21.

A SUMMARY

OF THE

REST OF THE CAMPAIGN.

29 Then Joshua passed from Makkedah, and all Israel with him, unto Libnah, and fought against Libnah: 2555 A. M. 1441 B. C.

30 And the LORD delivered it also, and the king thereof, into the hand of Israel: and he smote it with the edge of the sword, and all the souls that *were* therein; he let none remain in it: but did unto the king thereof as he did unto the king of Jericho.

31 ¶ And Joshua passed from Libnah, and all Israel with him, unto Lachish, and encamped against it, and fought against it:

32 And the LORD delivered Lachish into the hand of Israel, which took it on the second day, and smote it with the edge of the sword, and all the souls that *were* therein, according to all that he had done to Libnah.

33 ¶ Then Horam king of Gezer came up to help Lachish; and Joshua smote him and his people, until he had left him none remaining.

34 ¶ And from Lachish Joshua passed unto Eglon, and all Israel with him; and they encamped against it, and fought against it:

35 And they took it on that day, and smote it with the edge of the sword, and all the souls that *were* therein he utterly destroyed that day, according to all that he had done to Lachish.

36 And Joshua went up from Eglon, and all Israel with him, unto ᵉ Hebron; and they fought against it: c See Ch. 14. 13. & 15. 13. Judg. 1. 10.

37 And they took it, and smote it with the edge of the sword, and the king thereof, and all the cities

thereof, and all the souls that *were* therein; he left none remaining, according to all that he had done to Eglon; but destroyed it utterly, and all the souls that *were* therein.

38 ¶ And Joshua returned, and all Israel with him to ᶠDebir; and fought against it:

f See Ch. 15. 15. Judg. 1. 11.

39 And he took it, and the king thereof, and all the cities thereof; and they smote them with the edge of the sword, and utterly destroyed all the souls that *were* therein; he left none remaining: as he had done to Hebron, so he did to Debir, and to the king thereof; as he had done also to Libnah, and to her king.

40 ¶ So Joshua smote all the country of the hills, and of the south, and of the vale, and of the springs, and all their kings: he left none remaining, but utterly destroyed all that breathed, as the LORD God of Israel ᵍcommanded.

g Deut. 20. 16, 17.

41 And Joshua smote them from Kadesh-barnea even unto ʰGaza, ⁱand all the country of Goshen, even unto Gibeon.

h Gen. 10. 19. i Ch. 11. 16.

42 And all these kings and their land did Joshua take at one time, ᵏbecause the LORD God of Israel fought for Israel.

k ver. 14.

43 And Joshua returned, and all Israel with him, unto the camp of Gilgal.

THE SHADOW

ON

THE DIAL OF AHAZ.

Sharing the attention of the faithful, and equally demanding that of the merely scientific, stands the complement of Joshua's Long Day, namely, the absolute turning back of the sun in the time of Hezekiah. To reject one is to reject the other even more positively, and to accept either, logically demands the recognition of both. The latter is referred to three times in the Bible, and we quote each account at length.

ISAIAH.

CHAPTER XXXVIII.

1 *Hezekiah, having received a message of death, by prayer hath his life lengthened.* 8 *The sun goeth ten degrees backward, for a sign of that promise.* 9 *His song of thanksgiving.*

3293 A. M. 1,202,744. Day of the World, Wednesday, 18th, 1st Civil Mo., at High noon.

IN ᵃ those days was Hezekiah sick unto death. And Isaiah the prophet the son of Amoz came unto him, and said unto him, Thus saith the LORD, ᵇ³Set thine house in order: for thou shalt die, and not live.

2 Then Hezekiah turned his face toward the wall, and prayed unto the LORD,

a 2 Kin. 20. 1, &c.
2 Chr. 32. 24.

b 2 Sam. 17. 23.
3 Heb. *give charge concerning thy house.*

THE VOICE OF HISTORY.

<small>c Neh. 13. 14.</small>

3 And said, ^c Remember now, O LORD, I beseech thee, how I have walked before thee in truth and with a perfect heart, and have done *that which is* good in thy sight. And Hezekiah wept ⁴ sore.

<small>4 Heb. *with great weeping.*</small>

4 ¶ Then came the word of the LORD to Isaiah, saying,

5 Go, and say to Hezekiah, Thus saith the LORD, the God of David thy father, I have heard thy prayer, I have seen thy tears: behold, I will add unto thy days fifteen years.

<small>A Soli-Lunar Cycle.</small>

6 And I will deliver thee and this city out of the hand of the king of Assyria: and ^d I will defend this city.

<small>d Ch. 37. 35.</small>

7 And this *shall be* ^e a sign unto thee from the LORD, that the LORD will do this thing that he hath spoken;

<small>e 2 Kin. 20. 8. &c. Ch. 7. 11.</small>

8 Behold, I will bring again the shadow of the degrees, which is gone done in the ⁵ sun dial of Ahaz, ten degrees backward. So the sun returned ten degrees, by which degrees it was gone down.

<small>5 Heb. *degrees by,* or, *with the sun.*</small>

9 ¶ The writing of Hezekiah the king of Judah, when he had been sick, and was recovered of his sickness:

10 I said in the cutting off of my days, I shall go to the gates of the grave: I am deprived of the residue of my years.

11 I said, I shall not see the LORD, *even* the LORD, ^f in the land of the living: I shall behold man no more with the inhabitants of the world.

<small>f Ps. 27. 13. & 116. 9.</small>

12 ^g Mine age is departed, and is removed from me as a shepherd's tent: I have cut off like a weaver my life: he will cut me off ^t with pining sickness: from day *even* to night wilt thou make an end of me.

<small>g Job 7. 6.</small>

<small>6 Or, *from the thrum.*</small>

13 I reckoned till morning, *that,* as a lion, so will he break all my bones: from day *even* to night wilt thou make an end of me.

14 Like a crane *or* a swallow, so did I chatter: ʰ I did mourn as a dove: mine eyes fail *with looking* upward: O LORD, I am oppressed: ⁷undertake for me.

15 What shall I say? he hath both spoken unto me, and himself hath done *it:* I shall go softly all my years ⁱ in the bitterness of my soul.

16 O LORD, by these *things men* live, and in all these *things is* the life of my spirit: so wilt thou recover me, and make me to live.

17 Behold, ⁸for peace I had great bitterness: but ⁹thou hast in love to my soul *delivered it* from the pit of corruption: for thou hast cast all my sins behind thy back.

18 For ᵏthe grave cannot praise thee, death can *not* celebrate thee: they that go down into the pit cannot hope for thy truth.

19 The living, the living, he shall praise thee, as I *do* this day: ˡthe father to the children shall make known thy truth.

20 The LORD *was ready* to save me: therefore we will sing my songs to the stringed instruments all the days of our life in the house of the LORD.

21 For ᵐIsaiah had said, Let them take a lump of figs, and lay *it* for a plaister upon the boil, and he shall recover.

22 ⁿ Hezekiah also had said, What *is* the sign that I shall go up to the house of the LORD?

h Ch. 59. 11.

7 Or, *ease me.*

i Job 7. 11.
& 10. 1.

8 Or, *on my peace came great bitterness*
9 Heb. *thou hast loved my soul from the pit.*
k Ps. 6. 5.
& 30 9.
& 88. 11.
& 115. 17.
Eccles. 9. 10.

l Deut. 4. 9.
& 6. 7.
Ps. 78. 3, 4.

m 2 Kin. 20. 7.

n 2 Kin. 20. 9.

II. KINGS.

CHAPTER XX.

1 *Hezekiah, having received a message of death, by prayer hath his life lengthened.* 8 *The sun goeth ten degrees backward for a sign of that promise.* 12 *Berodach-baladan sending to visit Hezekiah, because of the wonder, hath notice of his treasures.* 14 *Isaiah understanding thereof foretelleth the Babylonian captivity.* 20 *Manasseh succeedeth Hezekiah.*

3293 A. M.
703 B. C.

THE VOICE OF HISTORY.

a 2 Chr. 22. 24, &c.
Is. 38. 1, &c.

3 Heb. *Give charge concerning thine house.*
2 Sam. 17. 23.

b Neh. 13. 22.

c Gen. 17. 1.
1 Kin. 3. 6.

2 Heb. *with a great weeping.*

3 Or, *city.*

d 1 Sam. 9. 16, & 10. 1.

e Ch. 19. 20.
Ps. 65. 2.

f Ps. 39. 12.
& 56. 8.

m Sabbath.

g Ch. 19. 34.

h Isai. 38. 21.

i See Judg. 6. 17; 37. 39.
Isai. 7. 11, 14.
& 39. 22.

k See Isai. 38. 7, 8.

Wednesday.

l See Josh. 10. 12, 14.
Isai. 38. 8.

4 Heb. *degrees.*

IN ªthose days was Hezekiah sick unto death. And the prophet Isaiah the son of Amoz came to him, and said unto him, Thus saith the LORD, ³Set thine house in order; for thou shalt die, and not live.

2 Then he turned his face to the wall, and prayed unto the LORD, saying,

3 I beseech thee, O LORD, ᵇremember now how I have ᶜwalked before thee in truth and with a perfect heart, and have done *that which is* good in thy sight. And Hezekiah wept ²sore.

4 And it came to pass, afore Isaiah was gone out into the middle ³court, that the word of the LORD came to him, saying,

5 Turn again, and tell Hezekiah ᵈthe captain of my people, Thus saith the LORD, the God of David thy Father, ᵉI have heard thy prayer, I have seen ᶠthy tears; behold, I will heal thee: on the ᵐthird day thou shalt go up unto the house of the LORD.

6 And I will add unto thy days fifteen years; and I will deliver thee and this city out of the hand of the king of Assyria; and ᵍI will defend this city for mine own sake, and for my servant David's sake.

7 And ʰIsaiah said, Take a lump of figs. And they took and laid *it* on the boil, and he recovered.

8 ¶ And Hezekiah said unto Isaiah, ⁱWhat *shall be* the sign that the LORD will heal me, and that I shall go up into the house of the LORD the third day?

9 And Isaiah said, ᵏThis sign shalt thou have of the LORD, that the LORD will do the thing that he hath spoken: shall the shadow go forward ten degrees, or go back ten degrees?

10 And Hezekiah answered, It is a light thing for the shadow to go down ten degrees: nay, but let the shadow return backward ten degrees.

11 And Isaiah the prophet cried unto the LORD: and ˡhe brought the shadow ten degrees backward, by which it had gone down in the ⁴dial of Ahaz.

II. CHRONICLES.

CHAPTER XXXII.

3293 A. M.

24 ¶ ^c In those days Hezekiah was sick to the death, and prayed unto the LORD: and he spake unto him, and he ⁶ gave him a sign.

25 But Hezekiah ^d rendered not again according to the benefit *done* unto him; for ^e his heart was lifted up: ^f therefore there was wrath upon him, and upon Judah and Jerusalem.

26 ^g Notwithstanding Hezekiah humbled himself for ⁷ the pride of his heart, *both* he and the inhabitants of Jerusalem, so that the wrath of the LORD came not upon them ^h in the days of Hezekiah.

c 2 Kin. 20. 1.
Isai. 38. 1.

6 Or, *wrought a miracle for him.*

d Ps. 116. 12.
e Ch. 26. 16.
Hab. 2. 4.
f Ch. 24. 18.

g Jer. 26. 18, 19.
7 Heb. *the lifting up.*

h 2 Kin. 20. 19.

Upon the basis of these four accounts of the two events, as true history, we have investigated them against the Cycles of the Heavens which still continue to score off human "times and seasons," and have found that they accord with these cycles, and are agreeable to *Chronology.* It is, therefore, next in order to premise our discussion by a succinct statement of the results arrived at by calculation.

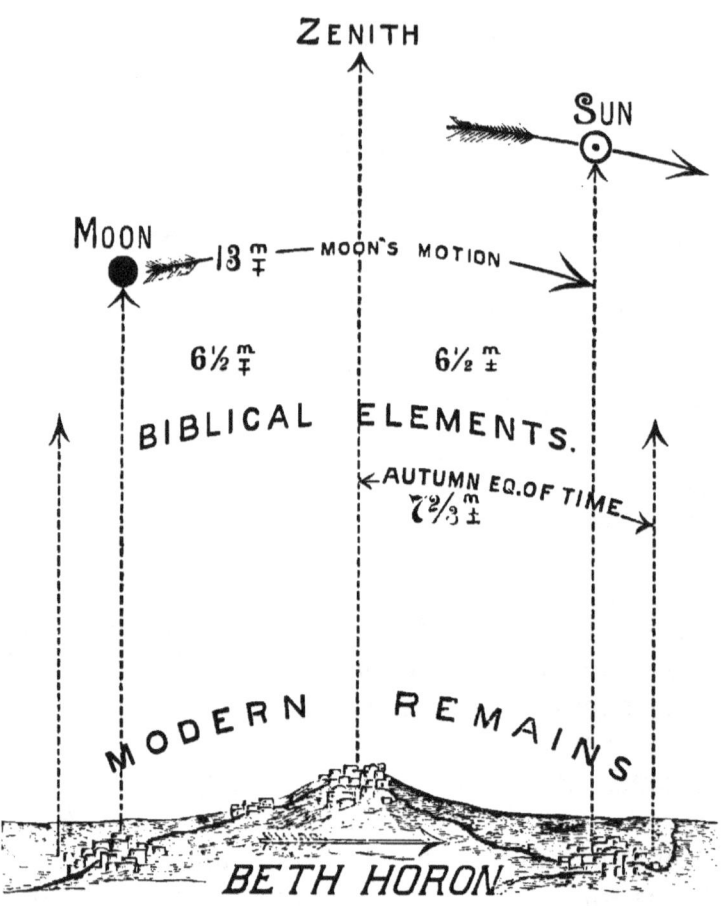

THE ELEMENTS VERIFIED.

It is of course impossible to give any adequate idea of the scope of the calculations which have conspired to bring out the astro-chronological results enumerated in this paper. The mere figures are of no interest save to the verifier; and even to him the eventual results will suggest far better ways of testing their accuracy than a mere going over of the tedious steps of the original and complex operation. If the results are correct, they must answer every other test which can be put upon them, and it is in perfect confidence that they will acquit themselves in this respect that they are now given to the world.

In the meantime, therefore, the chief point of interest to the scientific world is the explicit fixing of all the elements of the Beth Horon conjunction in modern terms, and from a modern starting point, as well as in Biblical terms and from a Biblical starting point, for the sake of the few who will be equally concerned to view it thence, and so for both to give the entire compass of the cycles which span human history.

To recapitulate, therefore, in anticipation of our discussion: Joshua's "Long Day" actually consisted of $23\frac{1}{3}$ hours added to the 24 regular hours which marked the day of the "winter solstice" of the year 2555 A. M., the autumnal-equinoxial beginning of which year was 3333 solar years ago, reckoning from Sunday, Sept. 22, 1889, A. D. These 47⅓ hours were considered as two full days by the calendar keepers of that time, and the single day which was, therefore, intercalated by them was more than, or chronologically *ahead* of the truth by 40 minutes; (a fact of which the Hebrews certainly seem to have preserved a careful record down to the days of Hezekiah, when, by an additional operation of Divine power, the calendar was set absolutely right). The days thus covered between the sunsets of the day in question were the 24th and 25th days of the 4th civil month of 2555 A. M., *i.e.*, the 113th and 114th days of that calendric year, and the 91st and 92d days after the 2555th completed solar year from Creation, dating from autumnal equinox to equinox, according to the universal method of all ancient nations down to Rome. They were also the last day of the 8th and the 1st day of the 9th lunation of the 2634th lunar year from this same epoch (Mosaic creation), or the 933,285-6th days of the world's duration, being respectively Tuesday and Wednesday.

Or, reckoning by reversed cycles from the solar-eclipsing new moon of Tuesday, June 17th, 1890, Joshua's "Long Day" was 3435 lunar years and 10 lunations *ago* (*i. e.*, 41,230 lunations, or from this same eclipse was 1,217,531-30 days ago).

THE BIBLICAL ACCOUNT. 19

The above mentioned last eclipse of History marked the 72,834th completed lunation of the moon, or the middle of its 6070th lunar year, or fell 4½ days short of our summer solstice of 1890 A. D. (*i. e.*, of the 5888¾th year of solar duration: or, finally, it took place upon the 2,150,-816th day of the world, and was the 22,862d eclipse which has occurred since the dawn of "Time."*

Finally, and in general terms, the Beth Horon conjunction was due 12-13 minutes past 11 A. M. on the first of the days identified, but, owing to the stoppage of all relative motion between "the three bodies" (and for aught we know throughout the entire universe!) was delayed "about a whole day" (23⅓ hours), and thus did not occur until 10.32-33 A. M. the next day, which was the Wednesday aforementioned; *i. e.*, the "silence" or intercalation, covered a part of both Tuesday and Wednesday, and the next sunset was the beginning of Thursday, the 933,287th day of the world.

The elements of the "Sun Dial" incident, during Hezekiah's reign, are as follows: It occurred at the absolute *instant* of Autumnal Equinox, in the year of the world 3293, *i. e.*, at the *end* of Astronomical year 3293 and at the beginning of Astronomical year 3294. The event took place practically at "high noon," *i. e.*, about 7½ minutes before 12 o'clock, as *we* reckon. The day was the

---
* *Vide* Appendix B.

1,202,744th from Creation, which was Wednesday, the 18th day of the 1st Civil month, of 3293 A. M., Ancient Hebrew Cycle. The moon, though involved in this incident, could not be mentioned, because she was just short of her entrance into her 4th quarter, therefore, at that moment (*i. e.* absolutely), was below the eastern horizon. There were no Palestinic landmarks by which to fix her place. That she, too, reversed her orbital motion cannot be denied, for her present place, (in arc), reverses through both this event and Joshua's, and strikes Creation's first hour without error. This could not have happened had she not been equally, and relatively, influenced in **Hezekiah's** day as well as in that of Joshua.

PART I.

JOSHUA'S LONG DAY,

AND

THE DIAL OF AHAZ.

THE DISCUSSION.

"And God said, Let there be lights in the firmament of the heaven to divide the day from the night; and let them be for signs and for seasons, and for days and years."   *Gen. i. 14.*

THE ARGUMENT.

"There was no day like that, before it or after it." Josh. x, 14.

The earnest seeker after Truth may enter her Temple by either one of several parallel corridors, which, upon examination, he will find to be joined by a sufficient number of cross passages to take him anywhere throughout the edifice. Along any one of these transepts he will find corresponding treasures, no matter in what main corridor he may be loitering. To mention but few: Astronomy, Chronology, Secular and Sacred History, lead, alike, directly onward to the altar. It is immaterial which corridor we select, but it is satisfactory to compare the vistas afforded along each, and thus obtain a better comprehension of the whole building so fitly joined together. The only object of this present guide-book is to point out some few of the lateral correspondencies in the Temple, and to demonstrate that they are parts of the same plan of architecture, in that they severally reflect their counterparts on either hand.

Upon June 21st, 1890, it was announced by the author that, as the final result of several years of close calculation, he had succeeded in identifying the conjunction of the sun and moon at which, "as it is written" in the Sacred record, "Joshua's Long Day" occurred,—and that his calculations completely vindicated the Biblical Chronology.

It is but natural that this announcement should have begotten a widespread comment, considerable misstatement, and awakened no little controversy and dispute; in view of which it is deemed wise to put the whole matter into as concise a shape as may be, without resorting to transcendental mathematics, which the average human being must—(and does in all calendric work)—take upon faith.

For instance, if almanac-makers had to supplement their work by an appendix giving all the figuring incidental to their finished tables, it would be a handicap sufficient to block their entire publication. But some will say, "the proof of the pudding is in the eating," and the safety of the modern almanac-maker consists in the fact, that the generation which uses it has constant demonstration of its accuracy. Very good, but the writer has almanacs and calendars in his possession extending back to the beginning of the century, almanacs of years before he was born. To him they are mere history, he cannot summon up his own experience. What then? Why, upon the foregoing premises, he is either forced to recalculate them, somehow or other, or else to accept them upon the basis of common sense, and to assume that they must have re-

THE BIBLICAL ACCOUNT.

corded true events, for otherwise they would have been repudiated by their own generation.

But this brings us to the gist of the argument in favor of the two events now under consideration. They occurred well within the days of written history, and were written down into the chronology of the days in question. They were consequently accepted in their own day, and must therefore have had foundation whereon to claim and effect an entrance into the contemporary history of generations who raised no voice against them,—they were admitted into their Chronology!

SECULAR CORROBORATION.

The writer does not pretend to explain *how* the Day in question was lengthened, but accepts it as a literal fact fully corroborated by history.

The Grecian Herodotus, whom we moderns call "The Father of History," verifies it by quoting the records shown to him by the priests while he was in Egypt. This is independent testimony, for neither the Greeks nor the Egyptians refer to the Hebrew account as the source of their own. But we may also refer to the Chinese, who preserve still another independent record of a similar event, (no doubt the identical one), as having occurred in the reign of Yeo, who was contemporary with Joshua.

In view of these four independent, and widely separated accounts, we must consider the event to be indisputable, so far as its *historical* evidence is concerned; and we maintain that no "*wise*-man" will say a word against the possibility of reducing the relative motion of the three bodies, (Earth, Sun and Moon), even to a standstill, until he really knows and can explain how that motion is produced!

No less eminent a philosopher than Newton has demonstrated how quickly the earth-motion might be slowed down, without appreciable shock to its denizens; and while the scientist can easily illustrate it, the devout astronomer,—(and "the undevout astronomer is mad")—can as easily offer a natural method whereby the stoppage could have been brought about.

For instance, any one may see, in the windows of the optician, a tiny apparatus, with four fans, which turn around with considerable speed, when exposed to the light. It is perhaps as near an approach to perpetual motion as man's ingenuity can make. If we merely raise our hand to interrupt the direct action of the actinism in the sun's rays, we immediately reduce the velocity of this apparatus by about one-half.

Now a cometary mass interposed between the Earth and Moon, and the Sun, in Joshua's day, might have easily cut off the actinic rays of the Sun, without affecting the light and heat rays of the spectrum at all, and so have accomplished the phenomenon.

How it really was accomplished, God only

INCEPTION OF PROBLEM. 27

knows; that it *was* done, Secular History testifies, and Sacred History asserts, with an authority not to be apologetically ignored by any one who prays sincerely that he may be "sealed" with saving faith.

It is an unfortunate sign of our times (2 Thess., ii. 3), that one who enters upon such a calculation as this, receives but little encouragement or response, even from quarters where he has a right to expect it, while in the scientific world at large the announcement seems to have begotten little else than disparaging discussion.

Nevertheless, a sufficiently large correspondence has already justified the belief that a publication of the main elements is now demanded, in order that they may be verified, or overthrown, by others.

THE INCEPTION OF THE PROBLEM.

The writer was led to make this calculation not to find out *whether* the account was true, but rather because he was convinced that it *was* so, and that it could therefore be PROVED to the satisfaction of any *reasonable* man.

And now it may be further stated, that in the course of his studies, and calculations, he has been led to see that the collateral fraction of an hour (40 minutes) required by the "turning back of the shadow upon the Dial of Ahaz," in the days of Hezekiah, also forms an integral, and necessary, part of human Chronology, and must be restored to it, in order to work the cycles of astronomy with absolute accuracy.

The fact is there can be no compromise in the position of a believer in the Bible. We freely grant this to the Agnostic, as the essential substance out of which the only bridge whereon we can hope to meet and draw up the preliminaries of an eternal truce, must be constructed, and we equally insist upon its employment by the Gnostic. That is, we fully recognize the *logic* of fundamental "common sense," and maintain, with all men of sound mind, that it can be fairly demanded of the Bible, that it shall square itself to its own record, when tried at the bar of the most transcendental astronomy, since by the premises of creation, and inspiration, their authorship is One.

At present, however, we are only dealing with Joshua's "Long Day," and are not yet called upon to divulge when, how, where, or why, the addition of a fraction of a single hour became astronomically necessary, was actually intercalated in the calendar, and was demanded in the calculations which verify the whole—all this in its proper place.

The problem, whose solution was primarily undertaken, was: whether a conjunction of the Sun and Moon occurred in the celestial vicinity of Beth Horon, at midday during Joshua's life.

The Chronological Conditions.

The chronological conditions imposed upon the problem by the sacred record, required that this mid-heaven conjunction should have occurred during the first five years of Joshua's occupation of the land, (2553–58 A. M.), and within these years the special Geographico-Astronomical conditions required that, by reversing the cycles of "the three bodies" from their *present* positions, their relative places should be such as to bring the Sun over Gibeon, and the Moon over Ajalon, within the set chronological limits. That is, the question to be settled was, whether Astronomy would corroborate History?

The battle of Beth Horon must have occurred during the first five years which succeeded the "passage of the Jordan". (Friday, 10th of 7th civil month, 2553 A. M.), and which preceded the "division of the land."

Caleb was 40 years old when sent out with the spies (Josh. xiv. 7), and was 85 when the land was divided (xiv. 10). Hence, the "division" was affected 5 years after their occupancy of it. For Caleb must have spent 40 of these remaining 45 years, in the wilderness with the rest of the host—*i. e.*, from the middle of 2513 A. M.—the date of the exodus—to the middle of 2553 A. M. —that of the entrance into Palestine.

The Battle of Beth Horon could not have occurred *after* the "division of the land," for long before that event we are expressly told (Josh. xi. 23), that "the land rested from war;" nor, of course, could the battle have occurred *before* the passage of the Jordan!

We are thus confined within very narrow chronological limits even before we undertake the crucial test of pure astronomy in order to find out the exact date which satisfies the rigid conditions directly imposed upon it by the record itself.

These conditions are as explicitly fixed by the account, as if it were committed, in so many words, to a "transit of Venus," upon the Beth Horon high-noon in question. Thus: in the account given in the Xth chapter of Joshua, we find the sun (☉) placed upon the meridian of Gibeon (35° 10′ $\pm$ E. of Greenwich), which latter place (Gibeon) lies east of Beth Horon by some 6′$+$ of arc; while, at the same time, the moon (●) is located upon the meridian of Ajalon (35° 2′ $\pm$ E. of Greenwich), at about an equal distance of arc (6′ $\pm$), to the *west* of Beth Horon.

The moon was therefore *recorded* as about 8′ $\mp$ west of the sun, and had the relative motion of "the three bodies" not been arrested, she would have come into conjunction (*i. e.*, become "*new*") in about thirteen minutes of time.

Now if the account is to maintain its integrity, —and Faith of course believes it will, while Infidelity never investigates at all, unless it be under a predisposition to find error,—we have here a most consummate set of astro-geographico-

AN ANOMALOUS REFERENCE.

chronological conditions, which must agree with the present positions of "the three bodies," to the very last degree of accuracy.

That they actually fulfill these complex conditions, and are rigidly true to the cycles as unrolled down to the date of the last sun-eclipsed conjunction of history (June 17, 1890), my own calculations verify to ix^{ths} (′′′′′′′′), or to 60^7 beyond seconds of time!

REFERENCE TO MOON ANOMALOUS.

But right here it is proper, and *apropos*, to insist that the mere mention of the moon, under the circumstances involved at Beth Horon, is a positive and *prima facie* guarantee of historical accuracy in the whole account.

For, as it was at *midday* that Joshua found himself at Beth Horon, and the moon, both by modern calculations, and by the tenor of the record, was so near to the sun (*i. e.*, at that portion of her orbit where she is always invisible even at night), there is no human probability that she would have been mentioned at all had not the facts of the case both warranted it, and demanded it, as a necessity. For about 27½ hours, both preceding and following a conjunction, the moon has no "phasis," and the Bible account places her within but fifteen minutes of the sun!

Bathed in such a meridian sun-glare she would have been invisible even to the Lick Telescope, and nothing but the veracity of the fact will ever reasonably account even for her incidental introduction into the record of this stupendous effort of the Solar System.

But *being* there, and being, moreover, an essential and ruling element of Hebrew Calendric methods, the whole system of Sacred Terrestrial Chronology demanded that she should be involved in the same mandate of "silence" imposed upon the sun, under penalty, if not, of throwing all the writings of Moses into unutterable confusion! For these writings are strung together, historically, in terms of a Lunar calendar, pure and simple, while at the same time their Chronology is consummately intercalated in order to keep solar time also. Hence to have held the sun, which did *not* rule the Hebrew "working" calendar, and to have suffered the orb that did so to pursue her "lost" way, would have necessitated an entire re-editing of the Pentateuch, or else have required all future generations to correct its chronology by the use of a "constant" of the most complex mathematical character.

Such a stoppage, therefore, as Joshua was led to request (Josh. x. 8.), demanded a stoppage of the moon as well as of the sun, *i. e.*, the earth's rotation, and the moon's orbital motion, had equally to be controlled. It is to this fact alone that we are indebted for any mention of the moon, and so her absolute place in her orbit is as positively fixed by the record, as is that of the sun,

A LOGICAL RESULT.

which was, after all, the ruling light, so far as merely Beth Horon's *battle* is concerned.

REAL EFFECT OF SUCH A STOPPAGE.

The effect of the stoppage of all relative motion among "the three bodies" for "about a whole day," was merely to introduce a single week-day into the calendar, and this was effected by the Hebrew priests, then and there, as a separate and distinct "measure" of the *duration* of the stoppage itself.

But, in so far as the actual measure of *celestial arc* is concerned, it could not, and did not, lengthen the then current year, 2,555 A. M., or lunar year (2634th), by anything whatsoever. That is, they, the year, the lunation, and the terrestrial rotation, were severally completed, when they were suffered to resume their "speech," at the very same points of the Zodiac which they would have reached had the incident not occurred.

If the power of Jehovah had enforced this "silence" on the spheres for a whole year instead of for a single day, the cycles themselves would bear no evidence thereof *to-day*, save only to mark, as now they do, the fact and date of the conjunction *at which* it was recorded to have occurred,—*i. e.*, to have begun and ended,—for the logical, and astronomical, carrying out of the mandate, requires no change of relative arc measurements while the "silence" continued.

But accurate "chronology" *would* have borne true evidence of the fact, and of its duration, if so be it should have been as well preserved by the Priests, (who were the calendar keepers), as they did preserve that of the single day which is recorded to have actually occurred.

For instance, a similar stoppage of all relative motion among the three bodies *in our day* would be very accurately measured by the chronometers of every Observatory upon earth, and by the watches now carried by almost everybody. And undoubtedly such a stoppage, as to its duration, would be measured and accounted for, in terms of week days, each of twenty-four hours, and be given its proper place as such in the current calendar.

But, save in the terms of such week-day periods, we would have, and could have, no other record of it, for the cycles, when resumed, would inevitably finish their respective courses at the identical points of the Zodiac they would have reached had no such relative stoppage been imposed. Thus all astronomical record of the stoppage would be lost so soon as motion was resumed, unless History and Chronology should have independently kept watch of it by Geographical references, and by Time, measured in some other way.

I maintain therefore both logically, and astronomically, and also as a chronologist, that the sole question which modern astronomy has to ask of "the three bodies," as now moving, and duly recorded by their elements in the best Ephemerides and Nautical Almanacs of the day, is whether such a *Conjunction*, as the record demands, is also demonstrated to have taken place at Beth Horon within the limits which are equally set forth by the account.

Beyond this, the yea or nay of astronomy cannot go one single element of *"arc,"* which is ITS only measure of "time"!

But right here accurate Chronology steps in, and her testimony has the casting and deciding vote, for if it shall be shown that, while the ecliptical points reached by "the bodies that rule the times and seasons" are the ones duly demanded by astronomically recorded time, while nevertheless the points reached in the Septenary sequence of the week days (*i. e.*, in the Calendar, which is the sole province of Chronology) are *ahead* of the astronomic ones by an amount just equal to the alleged duration of the stoppage as recorded by history, then the demonstration of the problem is complete and mathematical, and cannot be gainsaid in the least by sound reason.

It is thus manifest how beautifully History, Chronology and Astronomy stand related to each other in preserving the record of human "duration," and how consummately they may mutually assist each other, in defying those who would belittle the accuracy of the infinite and infallible Word as "it is written."

SOLI-LUNAR CYCLES.

But to continue; not before nor since "Joshua's Long Day" has there been a date which will harmonize the required relative positions of the Sun, Moon and Earth, as conditioned in the Sacred Record and reversed from their present relative places.

We are here limiting ourselves to the most transcendental accuracy, and are ignoring a numerous group of perpetually recurring approximations which, to all common purposes, bring about a repetition of a midday conjunction at a given place after 19; 651; 5859; etc. years, (*i. e.*, in one of the several soli-lunar cycles). But it is not enough to determine the least common multiple of a year and a lunation, in order to obtain an *accurate Soli-lunar* DIURNAL cycle, for the Earth's *own* rotation must be rigidly included in the calculation if we wish to tie the repetition geographically to the zenith of any particular place.

This necessitates the reduction of a Solar day, of a Lunation in Solar days, and of a mean Solar year, to their very ultimates, say at least to v^{ths} (′′′′′), the determination of the resulting least common multiple, and *its* reduction back to years.

If any one will undertake this simple, but tedious operation, he will find, as the writer has found, that the period required is much more than 23½ quintillion years! Nor short of this stupendous Eon can the conjunction sought, or in fact any other conjunction, *accurately* repeat itself, at the same place, in even its simplest solar, lunar, and terrestrial time elements.

With such figures in mind the search for Joshua's Long Day is certainly akin to hunting for a needle in the universe,—the which is manifestly so much the better in the cause of accuracy! Nevertheless there it is, and it results equally as well whether we reverse their cycles from the present positions of the three bodies, or work them forward from the soli-lunar conjunction at the instant of that autumnal equinox which marks the very first day of Adam's mundane chronology.

The True Origin of Time.

It is manifest that if we believed the Bible to be superlatively accurate the latter method would be the most natural one to pursue; for, not only would it pass through the Beth Horon conjunction, but, by producing the cycles onward, would inevitably land us at the places where the several bodies are now found. And this in fact, as above stated, is what occurs whichever way we work the problem, and hence the logical and irresistible conclusion is that the Biblical record is without error, and that henceforth we *may* assume the "Mosaic Era" as a natural astronomical *point d'appui* in all the calculations of Chronology.

When a novice has been conducted systematically through the windings of a labyrinth to its inner *Sanctum*, and has been fully initiated into the principles which govern the correct means of progress, it is manifest that he will ever after refer all localities therein not to some shifting point in the outlying circumference to which new paths are being constantly added, but will always start his measurements from the one fixed point which governs all the rest.

But it is admitted that the novice in our illustration has to be *conducted*, the first time, from the outside in. And so it is natural, in the confused state of Modern Chronology, to demand of one who maintains the accuracy of the Mosaic system, that he should trace *his* way backward to its origin, and take the initiates with him, in order to beget in them the confidence he has himself.

It is trusted that before the present volume has been laid aside a modicum of this confidence will have been instilled into its readers, and that they will thereafter see sufficient ground whereon to think from Adam down the stream of time, instead of upwards from some accidental generation of his everlasting posterity.

But to return to our topic:

The Intercalated Day.

To be scientifically correct it may therefore be stated that the Sun and Moon were going into accurate conjunction, in the mid-heavens over Beth Horon, (as recorded in Joshua), for the 31-,

604th time (since their primeval conjunction on the first day of Adam's first week of time), on the 24th day of the 4th Civil, or 10th Sacred month of the Hebrew calendric year 2555 A. M., which day was a *Tuesday* at 11.13 A. M., it being the 933,285th day of the world reckoning from Creation *inclusive*. Whereas, if we reverse the cycles from the latest solar-eclipsing conjunction of history,* to wit, that of Tuesday, June 17, 1890, they pass unerringly backward to that same conjunction, and make it 1,217,530 days "ago," but upon a *Wednesday*, at about 10.43 A. M.! *i. e., there is inevitably* "ABOUT A WHOLE DAY" *between the two results !*

Now, as to these intervening 23⅓ hours, Astronomy is *dumb*, and will be dumb FOREVER, while History—in Palestine, in Greece, in China, and in Egypt—is eloquent, and Chronology, in God's word, "is so written" that woe betide the fool who rushes in "where angels [and even devils] (Luke iv. 12-15) fear to tread."

This conjunction found the sun over Gibeon, the moon over Ajalon, and Joshua, in the height of battle, at Beth Horon, exactly midway between them. That is, the sun and moon were, to the last element of "arc," in Joshua's mid-heavens!

It is useless to contend against these figures, for they square with all the eclipses, transits, and equinoxes of Astronomy, and will land even a fair approximator at an epoch which will not

---

* *Vide* Appendix C.

repeat itself for a period of years whose aggregate is not to be counted by the whole human race laboring steadily thereat from Adam down to the present moment.

They also square themselves forward from Creation, with every date mentioned in the Holy Writ,—from Genesis to Revelations,—and backward, from the present, with all those of secular history, running back until they meet the former, and thereafter corroborating the sequence as one and the same thing.

Who then, in the face of them, shall arrogate unto the littleness of himself, so intimate a knowledge of the essence of celestial motion, as to dare to say that JEHOVAH, who hath wound the cycles up of old, did not also impose upon them such conditions as to bring about the event recorded in its proper day? Or who shall lift up his ephemeral "speech" against the "silence" which the common Maker imposed upon their's, because, forsooth, he cannot comprehend the universe from the stand-point of an earth-worm?

THE SITE UNIQUE.

It is at once noticeable, to an investigator of the geographical location of the places concerned in this incident, that the difference in longitude of Beth Horon and Gibeon, or of Ajalon and Beth Horon, so closely as modern Geography locates them, is equal to the autumnal "equation of time," while at the date of the conjunction itself, (winter solstice), there is *no* "equation of time;" that is, at this time of the solar year, mean and apparent time agree! In view of the surprising concert of Geographical, Astronomical, Historical, and Chronological elements involved in this chapter of Sacred History, it is astonishing to the writer that the eye of Science has never before been attracted to it, and that the mind of the devout believer has not long since seen in it the very site whereon to fight the decisive offenso-defensive battle of Faith against Infidelity.

Let us now describe the events at Beth Horon in the light shed upon them by the results of this tardy calculation.

Joshua crossed the Jordan on Friday, the 10th day of the 7th Civil (1st Sacred) month of the year 2553 A. M., and, without enumerating the intervening events, was in his permanent camp at Gilgal on Monday, the 23d day of the 4th Civil month of 2555 A. M.

This was at winter solstice, and sheds light upon the wisdom of the Amorites in selecting this occasion, as the most promising one, upon which to wipe out the only native allies of the Hebrews.

It also accounts for the fact of Joshua being found quietly camping with his hosts during the stirring task which had devolved upon them.

The Battle Described.

It was at this juncture that the men of Gibeon sent hastily to Joshua the news that they were surrounded, and besought his immediate assistance. There was no time to be lost, and Joshua's preparations seem to have been so quickly made as to have enabled him to leave Gilgal with the setting sun.

The sunset of his departure was of course the commencement of a day; and by calculation was on a *Tuesday's* commencement, according to the original, and then universal method of keeping the calendar. So Joshua marched all that night (Josh. x. 9), and, as armies move, reached Gibeon, some 20 miles away in the south-west, probably about dawn. The night was pitch dark, for the moon was going "new," and the surprise of the Amorites seems to have been complete.

The generalship of Moses and Joshua cannot be doubted, and the whole tenor of this particular account implies an adherence to strategical principles of the highest order by the latter. His first aim was to relieve Gibeon, his second to cut off retreat towards Jerusalem, and his third to drive the allies into the broken country. He was north-west of Gibeon when he started from Gilgal, but the account of their flight shows that the allies were forced to retreat along that very line! Joshua must therefore have made a wide detour to the south-east and have actually come upon them from their own flank and rear. We doubt if many modern armies would sustain such a surprise with equanimity, and it was certainly too much for the Amorites.

From the very first they were overcome with a great slaughter, which began at Gibeon, for true to the word of the Lord, which in spite of the urgency of his preparations Joshua had not failed to consult, he feared them not, knowing beforehand that they had been delivered into his hand, and that not a man of them could stand afore him (Josh. x. 8). Joshua of course fought with great odds in his favor, but certainly with no surer chances than any one may have who also has the God of Sabbaoth upon his banners.

Surprised, out-flanked, reversed in fact, and so cut off from their safest base of operations—Jerusalem, a city not wholly reduced until David's time—there was nothing left them but to seek individual safety in the wilderness. It was more than what soldiers call "panic" that dominated such a rout, for a forgotten God—the only God, and a God unknown to any but the seed of Abraham,—had stretched forth his arm and there was none to stay it.

Thus the Lord discomfited them before Israel, and slew them with a great slaughter, first at Gibeon, and as they fled by the way of Beth Horon,—a place some four or five miles to the north-west, and midway between Gibeon and Ajalon, which latter places were only 7 or 8 miles apart—the Lord of Hosts still pursued them, and smote them even to Azekah, and unto Makkedah.

Joshua, and his hosts, in the meanwhile closely profited by this supernatural assistance, and, following the retreating enemy (v. 10), the battle was probably at its height towards 11 o'clock A. M., and waging around Beth Horon.

That a severe convulsion of nature had already begun is manifest from the circumstances detailed in verse 11, where we learn that "it came to pass as they fled from before Israel, *and were going down to Beth Horon,* that the Lord cast down great stones (*aerolites?*) from heaven upon them, unto Azekah, and they died. There were more which died with hail-stones than whom the children of Israel slew with the sword."

The Long Day without Sunset and with no Sunrise!

By this time Joshua himself must have been in the vicinity of the elevated central point of the broadly extended battlefield, and the moment had arrived to announce the outcome of the prodigy which was already in progress.

The sun and moon, at this moment (11 A. M.) were absolutely in the "mid-heavens," equally distant to the east and west of Beth Horon—Joshua's own zenith—and about thirteen minutes of time apart, that is, they were, respectively, over the meridians of Gibeon, and Ajalon, to his right and left, as he pursued the enemy northward.

THE LONG DAY.

The conjunction was accurately due at the 17^h 12^m 56^i 48^{iii} 18^{iv} 16^v 47^{vi} 24^{vii} 26^{viii} 40^{ix} of that day, counting from its sunset beginning, say at our 13^m past 11 A. M. of Tuesday, December $21^{st.}$ It is at this juncture, therefore, that the incidents recorded in Joshua x, 12, 14, took place, and which, as we have seen, are now so circumstantially verified by History, Geography, Chronology, and Astronomy acting in concert. They are "written" as follows:

"Then spake Joshua to the Lord, in the day when the Lord delivered up the Amorites before the Children of Israel, and he said in the sight of all Israel, '*Sun, stand thou still upon Gibeon; and thou moon, in the valley of Ajalon.*'

"AND THE SUN STOOD STILL, AND THE MOON STAYED, until the people had avenged themselves upon their enemies.

"Is not this [also*] written in the book of Jasher?

"So the sun stood still IN THE MIDST OF HEAVEN, and hasted not to go down, ABOUT A WHOLE DAY.

"And there was no day like THAT before it, or after it, that the Lord harkened unto the voice of a man: For the Lord fought for Israel."

The Hebrew text states that the command to the Sun and Moon was "Be Silent!" and that the duration of this "Silence" was *about* a *whole* day," *i. e.*, $24 \mp$ hours.

---
* Vide Appendix. A.

It therefore covered the remaining part of Tuesday, and ran over (if *full* 24 hours) to about the corresponding hour of Wednesday, and thereafter, up to *that* sun-down, the remaining hours of Wednesday were completed.

It is now to be noticed that within the first 13 minutes which succeeded the resumption of relative motion, the *delayed* Conjunction must have taken place, and therefore that it occurred just where our modern reversion of the cycles demands, *i. e.*, upon a Wednesday agreeing with the very sequence of the week-days now kept in our modern Calendars, and 1,217,530 days *before* our Tuesday, June 17th, 1890.

The particular sun-down which succeeded this conjunction thus marked the Hebrew origin of Thursday, the 933,287th "day of the world," and sometime during it, and the next day, Friday. Joshua *may* (?) have returned to his winter camp at Gilgal in time to rest upon the Sabbath, which was the 28th day of the month.

Without knowing at all how the actual days of the *week* fell into the account, there has always heretofore. been more or less controversy over verse 15 in the account of this battle, where we are told that "Joshua returned and all Israel with him, into camp at Gilgal." A great many suppose that this indicates a *temporary* return, and that Gilgal was again left upon the receipt of news that the five kings were in a cave at Makkedah. As we have pointed out he *may* have done so, but from the *military* standpoint such a view is utterly untenable, and a more careful exegesis of the whole chapter bears us out in the conclusion that this was not at all the case. It is the opinion of the writer that Joshua was fortunate if his armies got back into their camp at all that winter, and at any rate that this particular Sabbath rest was spent at Makkedah, where his temporary camp was most naturally pitched during the closing hours of the "Long Day" under discussion.

A MILITARY COMMENTARY.

The Xth chapter of Joshua describes an entire campaign. In the first five verses we have a general account of the incidents which occasioned it, and their chronology sweeps from Israel's entrance into hither-Palestine, down to the winter when this particular campaign was undertaken. The next two verses refer to the day (Monday) on which the news reached Joshua. And, in order to allow him as many of its full 24 hours for his preparations, as the succeeding context requires, we may be sure that the messengers of the Gibeonites must have left their own city upon Sunday, or the first day of the week. We are here making a close, but none the less important, chronological point; for, by the time we have reviewed this chapter, it will be thereby patent that it fairly gives us, day by day, the incidents of an entire week, concerning whose central and most important day we now know all the chronologico-astronomical elements of "the three bodies."

Joshua's preparations were finished by Monday's sundown *termination*, and leaving that night (Tuesday), the events of the battle are generally described in the next 8 verses (8-15), particular prominence being given to its chief event,—*i. e.*, to God's manifestation of power in the sight of all Israel, of her particular enemies, and in fact of all the world, since we have independent testimony thereof from secular records. This closes the first sketch, as it were, and naturally ends with verse 15 as an outcome of the matter.

The 12 verses which follow (16-27) contain the special details of additional circumstances, and refer, in reality, to events connected with this same Long Day. The sacred historian follows the usual method of *raconteurs*, who, having given the main facts, return to special points and clear them up incidentally. A reference to the map will indicate the probable routes taken by the discomfited allies. The main body was undoubtedly driven via the two Beth-Horons over the mountains of Ephraim, and down by Ajalon; thence they sought concealment, each, in their own territory. But another column could have found a more direct gate of hope between the mountains of Ephraim and those of Judea. The troops of Jerusalem, however, with their king, were hopelessly cut off, and certainly took the Beth-Horon road; and it is manifest that the closely pursuing Israelites would have kept them in full view as they went down the western slopes of the mountains and turned towards Makkedah in the south. The fact, too, (16) that all the kings were eventually found hiding in a cave near this latter place, would imply that they had kept together, and had accompanied the main column. At any rate the two columns of refugees would have ultimately crossed each other near Makkedah. Here the confusion would have been still further increased, so that there remained nothing but concealment in that land of caves, so familiar in later days to David and his outlaws.

Joshua must have been in the vicinity when this place of concealment was discovered, and that the battle was still in its heat is settled by his commands recorded in verses 18 and 19. Verse 20 conducts the pursuit to its legitimate military termination, and from verse 21 we learn that Joshua himself had established his headquarters at Makkedah, probably from the time that it became of such special *tactical* importance.

The Battle was now over, and it is likely that, as the incidents described in verses 22-27 were begun, the sun and moon took up their accustomed motions. There were about 7¼ hours remaining to the day, and as it closed (v. 27) the bodies of the dead kings were taken down and cast back into the cave. It was of course Thursday "evening" (*i. e.* its beginning) when this final work was completed. The night of this new day was certainly spent in much needed rest, but the *latter* half of its Hebrew duration (its "morning") is plainly referred to in verse 28 as *the* day on which the city of Makkedah fell. The next day (Friday) was the preparation for the Sabbath, both of which latter days, in view of all the circumstances of the case, were undoubtedly spent in Camp Makkedah,—and so closed this most remarkable week.

Beyond this point we cannot follow the matter by dates, nor is there any necessity for so doing. The rest of the chapter merely gives the broad outline of the general campaign which followed, and which spread from Gibeon in the north, to Gaza, indeed, into Goshen itself, upon the very borders of Egypt, and thence due east to Kadesh-barnea (41). That it was continuous is shown by verse 42, and by the carefully recorded sequence between its six sieges (Makkedah, Libnah, Lachish, Eglon, Hebron, and Debir, to say nothing of innumerable minor cities implied in the account) and its one pitched engagement (verse 33); that it was relentless, and for good cause, we may be sure from verse 40, and that not until it was completely finished (probably not before Spring was well advanced) did Joshua return to Gilgal, is settled by verse 43.

The Conquest of Palestine.

Thus ended the second of Israel's campaigns for the conquest of the land upon "this side of the Jordan," nor can we refrain from pointing out the consummate generalship with which the three campaigns—that of the Center for Samaria, of the South for the Amorite country, and of the North for Galilee,—are strategically united.

By passing up the eastern side of Palestine until opposite Jericho, the land was entered in what from the military stand-point we term its "middle zone." Here the decisive battle of Jericho, and the eventually successful campaign against Ai, struck such a sudden and stunning blow against the inhabitants that for a long time the now separated Northern and Southern nations hesitated to meet Joshua in open contest. The Gibeonites obtained immunity by a cunning stratagem, and en route through their territory, from the first and original Camp Gilgal (Josh. iv. 19), the host moved north to that Gilgal which is in the land of Ephraim, and which became their second and really "permanent camp."

The new year 2554 A. M. had now begun, and its fall was spent in preparations for the first winter in the land. The spring of this year undoubtedly found Joshua busily engaged in the operations around Mt. Ebal (Josh. viii. 30–35) quite near to his permanent camp. Here he caused the great altar to be erected out of whole stones whereon no man had ever raised a chisel, and thereafter the entire law of Moses was hewn into it. This undertaking must have consumed a good part of the rest of that year, and in view of the care with which it was done, a careful archæological examination of Mt. Ebal might repay modern research and exploration far better than the costly excavations in the ruins of Babylon.

This undertaking was not only accomplished in the presence of all Israel, but when completed was followed by a celebrated feast, no doubt the New Year feast of 2555 A. M., at which every word that Moses had ever commanded was read in the ears of all concerned.

Israel now returned to Gilgal, spent the Fall in preparation for their second winter, and at its solstice became involved in the campaign we have just examined.

It is not to be understood that the Amorites, in so far as they were concerned, entered upon this undertaking without long and careful preparation. From the very passage of the Jordan they had become demoralized (Josh. v. i), the fall of Jericho had vastly increased the fame of Joshua throughout all the country, (vi. 27), and the destruction of Ai, with which the opening campaign ended, followed by the defection of the Gibeonites, manifestly demanded caution and we need not doubt begot it. Nevertheless, their arrangements, conceived from the first (ix. 1-2), at

CONQUEST OF PALESTINE. 53

last took form (x. 1-5), and met the fate which we have sketched.

To his military position between the northern and southern peoples of the land is no doubt due the fact that none of the former were included in this effort; not indeed that it would have been any more successful, but none the less from every stand-point Joshua's generalship is exalted.

Joshua occupied Palestine as Napoleon did the field of Austerlitz, and having now swept its center and south, save the city of Jerusalem, and the land of the Gibeonites, was in the most advantageous situation to accept the challenge of the northern kings.

His campaign against them is described in the next chapter, xi, and was by far the most prolonged of the three (verse 18), consuming probably at least two and a half of the three remaining years, and at their termination "the land rested from war" for the remainder of his days.

The land was thus conquered "in detail," and from the "center outwards," nor has modern warfare any fault to find with his fundamental military principles.

CORROBORATED BY ECLIPSES.

But to return again to the discussion of Joshua's Long Day, and of the true chronology thence resulting. Of course, the announcement of this calculation has awakened criticism and dispute. For instance: It has been pertinently asked, why "if this line of time is now at last so accurately defined, does not the professor make it pass through some well known date, or, better yet, string an eclipse or so upon it, so that we, who prefer to feel our way carefully backward from the present rather than boldly start out from Adam's era of 'Paradise,' may at least have some one of these lunation mile-stones marked with its true and intelligible elements? Until this is satisfactorily done we shall continue to be in doubt as to whether even this new moon conjunction (to say nothing of its extra astronomical pretensions) is a 'fake' or a 'fact.'"

The position taken by this correspondent is the natural, modern and scientific one and I should be certainly "in the vocative" if I could not furnish astronomical waymarks, backward from the present, so well as forward from creation's dawn, whereby to verify my work.

It is as manifest to me, as it can be to a "practical astronomer," that if my line of lunations is correct I should be able to identify upon it some of the eclipses of the past. This is exactly what I can do, and in fact what I have already done in order to verify my calculations to my own satisfaction, and with this result: That the line being correct it serves to identify every eclipse both of sun and moon that has ever occurred, or ever can occur so long as the solar system obeys the laws that now govern it.

This is not so extraordinary a claim as it seems. All solar eclipses must occur at new moon, and all lunar ones at full moon. There are at a maximum but 70 possible eclipses in a "Team" or sequence of 18 years and 10 to 11 days, after which they repeat in exactly the same sequence, and so move down the ages as unerring sign-posts.

These sign-posts are planted by the moon in her lunations, and if we have her line correctly, we can certainly identify not only any eclipse of History, but all the rest, whether recorded or not. Now to identify a single one, recorded some years ago, is to demonstrate to a "practical astronomer" that the line is right, and this I shall proceed to do.

Let us therefore refer to my original announcement, made in the New Haven *Register*, of June 21st, 1890, but by printers' mishap somewhat disarranged. The proper announcement was that 802 lunar years and 2 lunations ago, Joshua's Beth Horon conjunction was repeated, *i. e.*, reoccurred in due mathematical relation to the zenith of the same place.

As already noted, one must always speak advisedly as to repetitions, and with a full knowledge that they are merely approximations at best, and are of value to almanac makers only according to their degree. Our modern almanac makers are generally content with accuracy to days or hours, or at most to minutes, rarely to seconds, but the universe exhausts the very ultimate.

Now the *re*-conjunction to which I then referred, (aside from any of its merely local, or Beth Horon concomitants, to which however it was duly, and mathematically related), took place upon March 29th, 1112, A. D., and as I announced in the New Haven *Register*, of July 2d, 1890, it was then *additionally* marked by a solar eclipse, to-wit No. 52 of the Regular Team. There was no eclipse, but simply a conjunction, upon the Day of Beth Horon, but in 1112 A. D., the eclipse referred to was brought about by the fullness of *other* cycles.

This eclipse of March 29, 1112 A. D. was followed the next month (April 13), by a lunar eclipse (No. 53), and at autumnal equinox of that year (September 22-3), the sun was again eclipsed *circa* the first points of Libra (No. 54). This latter eclipse was followed the next month (October 6) by another lunar eclipse (No. 55), and on March 18, 1113 A. D. (*i. e.*, exactly 12 lunations, or one full lunar year, after the one with which we started) at the 19th hour of the day, a full solar eclipse (No. 56) occurred at JERUSALEM, *and was then and there recorded*, as will be found by consulting the records of eclipses. Beth Horon, Ajalon and Gibeon are but a few miles north-west of Jerusalem, and an eclipse which involved one would almost certainly have compassed the others. This was in the reign of Henry I. of England, and of Baldwin I. of the "Latin Kingdom of Jerusalem."

Here, then, we have tied ourselves to an eclipse which is actually *recorded*, and have *thus verified the unerring* accuracy of our "line of time." Let me here state further that this latter eclipse (No. 56) will be repeated on April 16th, 1893, A. D., as a necessary resultant of the very same celestial mechanism, and may be predicted far more certainly than we can count upon the future chimes of any earthly chronometer—upon that day the sun will set, eclipsed, at Jerusalem.

To return now to our eclipse No. 52, which we stated to have specially marked the "repetition" of the Beth Horon conjunction of Joshua's day. The distance apart of these two new moons is exactly 31,604 lunations, or 2,555¼ solar years, and the first one, or Joshua's, actually occurred in the year 2555 A. M., at its winter solstice.

But as it is fairly demanded that the trace shall be *backward* from the present, rather than forward from an origin of time which is under dispute, let us take the new moon of this present month, now shining, and full to-day, (Wednesday, July 2d, 1890), which by the way is the central day of the current solar year, as we moderns fix it, in the A. D. system, and which may be still further anchored to the rigid facts of the solar system by noting that this "fullness" occurs with the sun in Apogee, and the moon in Perigee.

It is thus manifest that we could not have a better, nor more remarkable lunation from which to reverse our cycles, and feel backward to that far greater one which is the object of our studies.

This modern lunation under consideration, was *renewed* upon June 17th, 1890; reckoning, then from it, 3,435 lunar years, and 10 lunations ago, marks the conjunction of Joshua's Long Day.

In other words, the time backward is 41,230 lunations, no more and no less, and they pass through every eclipse both in history and out of history, because they start with the very last eclipse of history, to-wit: the annular one of the sun, which occurred on this very June 17, 1890, (our "starting point"),* and pass through the group mentioned in 1112 and 1113 A. D.!

In the humble opinion of the writer, this calculation has come to stay, and some day to be recognized at its full and intrinsic value, and he is confident that if he could calculate the trajectory of a human projectile as unerringly as he can rely on the motions of those that Jehovah has placed in the Heavens for "signs and for seasons," in so far as Adam's race is concerned, he would feel very little anxiety for America in time of foreign war even if he had to fight her battles single-handed. In the meantime it may be maintained that while the Bible needs no human bolstering to support its infinite accuracy, nevertheless we are constrained to believe that the human understanding itself *does* need such helps as this and similar calculations, in order to

---
* *Vide* Appendix C

force it back upon the sometime inevitable plane of implicit faith.

VERIFIED BY THE EQUINOXES.

As one among several other independent verifications of this calculation it is to be noted that the autumnal Equinox last year, 1889, was the $5,888^{th}$ since creation, and that it occurred upon the $2,150,548^{th}$, "day of the world," to-wit: upon Sunday, September 22d, 1889, as we know from the government ephemeris of that year, and which Sunday, in spite of Parliamentary enactments as to Greenwich mean noon, etc., did not really commence until its own modern sundown had been duly recorded by nature at the far-off eastern "*primary* meridian."

The ephemeris' time of this autumnal equinox was *circa* $8^h 45^m$ after Greenwich mean noon of that modern Sunday, which is set back from the ancient origin of day by just six hours.

Thus the true time was *circa* 2^h and 45^m after the Greenwich mean sunset *beginning* of this particular Sunday, or $9^h\ 37^m\ 31^s$, etc. after its sunset commencement at the *most ancient* "primary meridian."

Nevertheless it can come so by no possible mathematics without the interpolation, or "intercalation" of exactly 24 hours.

This intercalation is demanded by the $23⅓$ $\pm$ hours ("*about* a *whole* day") which compass the stoppage of relative motion upon Joshua's

Tuesday-Wednesday, together with Hezekiah's 40 $^{mts.}$ or ⅔ of a *single* hour (*i. e.*, 10° backward of sun motion), by which the calendar was finally set in absolute order.

All this is proved by the simple inspection and comparison, of the two following equations:

$$(a) \quad \frac{5888 \text{ Y}}{7} = 307{,}220 \text{ weeks } 6^d\ 9^h\ 37^m\ 31^s \text{ etc.}$$

which brings us (from the original Sunday) only to the 9th hour of a seventh or Sabbath day (to-wit: that of the 2,150,547th), and

$$(b) \quad \frac{5888 \text{ Y} + \left(23\frac{1}{3} + ^h_{\text{Joshua's}}\right) + \left(\frac{2}{3} + ^{}_{\substack{\text{Heze-}\\\text{kiah's}}}\right)}{7} = 307{,}221^{\text{wks.}}\ 9^h\ 37^m\ 31$$

in which latter equation the 9^h 37^m 31^s, etc., fall *where they actually came*,—as at creation,— namely, upon the requisite Sunday or a "first day of the week," Sept. 22, 1889.

This is the *dictum* of the modern ephemeris, and it is tied to every chronological element in the whole solar system although the latter consists of more than 250 intimately interlaced cycle-making orbs not one of which can be impugned without the condemnation of all the rest.

In the foregoing equations, Y is the mean-solar-year-value, and cannot now be altered 1 second plus or minus from 365^d 5^h 48^m 50^s 53$^{''}$ and 60^{iv}, while as to the number of years involved, the rigid work of the "British Chronological Society" has demonstrated, by the verification of all the eclipses and transits, both in and out of history, that the number of years spanned from the dawn of Genesis to our September 22d, 1889, is no more, and no less, than 5,888 of mean astronomical duration.*

SHADOWED ON THE DIAL OF AHAZ.

It is the firm conviction of the writer, fully borne out by certain conditions impressed upon the verification of these events as part of one grand entirety, that the actual duration of the stoppage of relative motion, in Joshua's day, was exactly 23⅓rd hours, and that, to avoid calendric confusion, the High Priest, or official timekeeper naturally authorized the intercalation of a full day (24 hours) at the time of the Beth Horon occurrence: that, nevertheless, it was always thereafter a matter of the most careful record that this intercalation was 40 minutes in excess of the truth.

This knowledge must have descended to the days of Hezekiah and Isaiah, the latter of whom, probably fully informed thereon, made double purpose in his later and equally extraordinary request that this remaining part of the missing hour might be, then and there, made up, and the Calendar thus made absolutely correct.

For, while ix^{ths} of time have not escaped me in this verification, I can find no indication of any

---

* *Vide* Appendix D.

calendric change as incident upon Hezekiah's request. Nevertheless, the totality of time between the primeval autumnal conjunction in "Eden" and the 72,834th, which occurred on June 17th, 1890, demands exactly 24 hours' interpolation or intercalation, *for the two events.*

All this is also borne out by a fair and critical examination of the texts concerned; (*vide,* and compare 2 Kings xx. 1–11; Isa. xxxviii; 2 Chr. xxxii. 24; and Josh. x. 13; etc., page 5).

Hezekiah ascended the throne at the end of 3278 A. M., and died at the end of 3307 A. M., having reigned exactly 29 years, the last 15 being from Equinox to Equinox. The Dial incident occurred at the beginning, or autumnal Equinox of the Solar year 3293 A. M.. *i. e..* just 15 exact Solar years before his death.

Isaiah's visit to him was upon Wednesday the 1,202,744th "day of the world." This was the 18th day of the 1st Civil month of 3293 A. M., and at the 18th hour thereof (or at "High Noon" reckoning from sundown), the Sun went into Autumnal Equinox.

It was at this very moment, and before Isaiah left the sick King's bedside, to which he had just previously *returned* (2 Kings xx. 4, 5), that, at the prayer of the Prophet, the "Shadow went back" 10°, or 40 minutes, "upon the Dial of Ahaz."

The sign was thus given at once, and upon the actual Solar New Year's *day* and *instant,* al-

SHADOWED ON THE DIAL. 63

though from the position of the then current year upon the Calendar the event has until now been completely hidden from us moderns.

There is no reason to doubt that the prodigy was quickly reported to the anxious monarch, by the High Priest of the day, for this latter, as the official Calendar keeper, would at that very moment have been closely watching the Shadow in the court without, and would have been so doing entirely unconscious of what was taking place at the same time beside the sick bed in the Palace, since he would, by mere virtue of his office, have been necessarily and *personally at the Dial awaiting the New Year instant!* It was at such a moment that "Isaiah the prophet cried unto the Lord; and He brought the shadow ten degrees backward by which it had gone down in the Dial of Ahaz."

The significance of these closely related circumstances is not to be underrated, nor should we lose sight of the fact that we are now, for the first time, sufficiently informed upon them to properly understand the *rationale* of what occurred upon this momentous day.

The Equinox of the year in question was a remarkable one at Jerusalem just because it occurred at local high-noon. The normal advent of this particular Equinox was undoubtedly foreknown as an astronomical event, and eagerly anticipated by those skillful star gazers. The preparations for its accurate observation were made beforehand as certainly as in modern days they would have been to watch a pre-calculated "transit of Venus." Moreover the machinery for all this observation already existed in a most elaborate form.

Ahaz, more than any of the kings of Judah, had turned his attention to Sabaism or star worship (2 Chron. xxviii), had erected its astrological altars throughout the city of his fathers, and he had copied the design of one, in particular, whose steps formed the famous Dial, from an original seen by him in Damascus, where he went to meet Tiglath Pileser (2 Kings xvi). This altar was placed right in the center of the Temple area; not only therefore was it conveniently located for meridional observations, but from that area the access into the King's Palace was direct.

We can thus easily picture to ourselves the two groups who were chiefly concerned in the event—the High Priest, with his attendants, carefully watching the Shadow as it moved slowly towards the Noon-mark, and the King, dying from a carbuncle in its last stages, doubtfully listening to Isaiah as he promised him so speedy a recovery, that in three brief days he could go out and pay his vows in person at the altar. These circumstances also lend peculiar light to the "sign" the prophet then and there suggested in verification of his message. What was going on without in the Temple's court was a matter of general information. The King, of course, knew it; Isaiah knew it; all Jerusalem knew it; and this very fact may have suggested to Isaiah the peculiar fitness of this particular sign under the circumstances. It was already noon, and the Shadow was probably just about to fall into coincidence with the meridian.

"Shall the shadow go forward ten degrees, or go back ten degrees?" now asked the prophet.

"And Hezekiah answered, It is an easy thing for the shadow to go down ten degrees," *i. e.* to pursue its course :—"Nay, but let the shadow turn backward ten degrees."

In the court without the intent group are just about to announce the meridional coincidence, when lo, the shadow suddenly reverses its easy motion, and an unexpected and unprecedented prodigy occurs. The shadow moves suddenly and steadily backward over a large section of the dial, and stands at the 20 minute mark. Forty minutes yet to noon! There is no doubting the evidence of one's own senses, and chief among those who were amazed stood the High Priest himself!

The surprise and consternation of this functionary can be judged of in a small degree *by the sense of awe* with which the present discovery of this so unlooked-for accuracy, and the fitness of the several phases of the incident, must strike the modern mind!

"The third day," (2 Kings xx. 5. 8) from this Solar New Year Wednesday-noon, of course brings us to the SABBATH DAY, the 21st of that current month, and on it the now fully recovered king most fittingly went up to "the House of the Lord" to render his doubly appropriate thanksgiving (Isa. xxxviii. 9-22).

The moon herself was equally affected upon this occasion, *i. e.* the stoppage, or actual reversion this time, was a relative one of the whole luni-terrestrial system.

But the moon is not mentioned in the account, because she was just short of her 4th Quarter (*i. e.* 21.94 days old), and, as it was "high noon," she was of course more than 90° away, *i. e.*, just below the eastern horizon, and, therefore, no Palestinic landmarks could be cited in her behalf. Nevertheless, as all astronomers know, the earth and moon are so rigidly related to each other in their dominant cycle (as if a steel bar joined their centers), that in this particular case she is as clearly implied in the *ipso facto* as if her actual position could have been geographically fixed.

It is questioned in the mind of the writer to which of these two stupendous events in the solar system he should accord the superior place. Perhaps the answer can never be fully satisfactory. The fact is they form the complements of each other, and have written into human chronology, by their combined action, a single day, unique among all others, in that it begins and ends in the "mid-heavens," and works from its commencement to its close without a setting or a rising sun!

Proved from the Almanac.

But perhaps the simplest calendric proof of the accuracy of the result of the main calculation discussed in this brochure, *i. e.*, the verification of Joshua's Long Day, is the following, based upon the Lunar or Metonic cycle, a period of 19 tropical years, or 19 years 2 hours and 3 or 4 minutes when the same moon occurs in her 235th lunation.

This period is a familiar one to all almanac makers, as well as to all who are versed in ecclesiastical or lunar chronology.

Upon it depends the age of the moon, or its "epact," which will be found recorded in every good almanac.

By this word "epact" we mean "the age of the moon" at the beginning of the year under consideration, and this depends upon the position of the year itself in the current Metonic cycle. Thus, in almanacs for this year (1890 A. D.) the epact is 9, which means that the moon was 9 days old on the first day of January. In other words, we are in the 10th year of the current Metonic or soli-lunar cycle.

Last year (1889 A. D.) was, therefore, a 9th year in this cycle, and its "epact" was 28.

Now, to apply this cycle to the case in point. *i. e.*, to the verification of the lunation which marked Joshua's "Long Day," it is to be carefully noted that the results of the calculation assert that the sun and moon were in accurate conjunction at mid-day, of the *winter solstice* of 2555 A. M., *i. e.*, their "epact" was at that time 0, or in other words a cycle was then beginning. If so, the age of the moon at the *winter solstice* of 5888 A. M., or the *beginning* of the 3334th year thereafter, as determined by the cycle, should agree with our almanacs of 1889 A. D. This is exactly the case, $3334 \div 19 = 175\frac{9}{19}$, *i. e.*, there are in 3334 years just 175 full cycles, and $\frac{9}{19}$ths of a cycle. Hence the solar year of duration, beginning at the winter solstice of 1889 A. D., was the 9th year of the 176th Metonic cycle from Joshua's Long Day. Therefore, the age of the moon at that time should have been as required by the following table, giving the "epact" corresponding to each of the several subordinate years:—

The epact is therefore 28, which will likewise be found to be the "age of the moon" upon the winter solstice (Dec. 21st) of 1889.

Any one may verify this by consulting a last year's Almanac or Ephemeris. But it may be proved from an almanac of the current year (1890), by noting the following which has already been alluded to.

The "epact" this year is 9, therefore the year is No. 10 in the cycle, hence last year, 1889, was the 9th year in the cycle, consequently its "epact," from the opposite table, was 28. But the "win-

THE 176th METONIC CYCLE
FROM JOSHUA'S LONG DAY.

A. D. Years.	Age of Year in the Cycle.	Corresponding Epact or Age of Moon.
1881	1	0
1882	2	11
1883	3	22
1884	4	3
1885	5	14
1886	6	25
1887	7	6
1888	8	17
1889	**9**	**28**
1890	10	9
1891	11	20
1892	12	1
1893	13	12
1894	14	23
1895	15	4
1896	16	15
1897	17	26
1898	18	7
1899	19	18

ter solstice" of our modern "common" years falls exactly one lunar year (354 days), after the beginning of the solar year January 1st, hence the age of the moon, upon both dates, was approximately the same, *i. e.*, 28 days, and therefore its age at winter solstice in 2555 A. M. was 0—*i. e.*, the moon was *new*, or in conjunction.

Now there is no astronomic loop-hole here, through which to escape the Q. E. D. of this result, for as the age of the moon at winter solstice of 1889 A. D., was certainly 28 days, the year of the cycle was as certainly a 9th, and as the epact of the moon at winter solstice of the current year (1890 A. D.) is certainly 9, the year itself is a 10th year upon the Era dating from the winter solstice of 2555 A. M. And, finally, the cycle we are now in must be the 176th from the winter solstice which marked Joshua's Beth Horon conjunction, because, otherwise, every eclipse of history would be thrown out of place, as is manifest by a reference to the single group which we have already identified.

The fact is, the beauty and accuracy of Biblical Chronology, when rightly understood, is beyond the compass of human language. It absolutely exhausts our finite means of numerical expression, and dwarfs the ken of even the sublimest mortal intellect.

To the minds of the faithful the foregoing discoveries and demonstrations will be welcomed with the keenest satisfaction, and, in that they redound unto the glory of Jehovah, they will gladly join the writer in the additional homage which they cannot but beget towards Him who is again shown by these humble efforts to be faithful and true forever and forever.

It has thus been permitted, to this almost already faithless, and certainly final, generation of the present dispensation, to thrust its hands into the *wounds of time*, in order that perchance they might thereby recover their integrity, and accept the literal Word of God; yet none the less will the former generations ever be more blessed, in that they "have not seen, and yet have believed."

THE APOLOGISTS ARRAIGNED.

And this brings us to a point where we can once again, and with the most feeling emphasis, condemn that particular and blasphemous phase of the so-called "higher criticism," upon whose shoulders so much of the infidelity *within* the fold can certainly be laid.

It is nothing short of scandalous to the Christian Church that it has so long permitted the very highest seats among the teachers of our generation, and the loftiest pulpits, to be filled by men who openly teach disbelief, which is "infidelity," in the grandest chapters of the Bible, and who, by their continued and promiscuous apologies to the enemy (2 Thess. ii), and their absolutely unwarranted and frivolous analysis of the Infinite Word, seek to make it of none effect, and so take all its saving power away from those who otherwise would gladly hear, and having heard, obey.

It is certainly imperative, in that now it is in fact the last few moments of the dread "half-hour of silence" which succeeds the Opening of the Seventh Seal (Rev. viii. 1), for men to break away from those who demonstrate that they have naught to sell, and hasten back unto "Moses and the Prophets," if perchance there may be time, yet, to replenish their lamps with oil, so soon and certainly to be sorely needed!

What the church of these starving "latter days" most needs, is a pure Biblical exegesis, founded upon explicit faith in all that the Word contains. We need an honest *explanation* of the Bible, and it is high time to devote what little there is left unto the teachings of the Prophets.

The Sword of Damocles.

It is now a full generation since our public services were conducted as if an "Advent" were not only promised, *but was imminent.* In the meantime—while we have listened to some of the "doctors" in the pulpit, as they have plumed themselves with divers and delusive theories, which not only have no power to save, but literally damn the soul of the believer, in that they necessarily engender the most insidious forms of infidelity—in the meanwhile the "*time of the* END" has literally crept upon us unawares,—"for the coming of the Lord draweth nigh," and behold the Judge standeth before the door!"

The whole tenor of the Scriptures points towards a calaclysm at the very height of what the world will consider to be the noon of promise. It looks towards the very state of affairs which now surrounds us, and is in fact the ozone of the air we breathe. Born into a delusion, we are blind to the reality, and, therefore, even as predicted, we are now overtaken in the midst of what we deemed to be the dawn of an eternal progress.

But the Prophets have not spoken in vain, for the faithful have, to the very limit of their light, paid heed unto their words, while the blind, though having eyes, have dwelt among these same scenes with such utter unconcern, that now they have no ground wherein to plant a single potential blade of wheat.

That the world is about to go into the most acute crisis of all history, may be gathered by any mind capable of generalizing among the diverse testimonies that surround us. Nevertheless none but "the wise" can possibly perceive the import of this truth.

Take even that Epistle of James, which Luther called an "Epistle of Straw," but which a wiser man, wiser in the light of modern tendencies, can perceive to be a brick filled with straw, and therefore bonded with endurance, and let any man, concerned with the problems of the day, and convinced as to their outcome, which is clearly set forth by Paul in 2 Thessalonians Chapter ii, read the fifth chapter of this straw epistle. He will see in it the same gospel, the same truth, the same inevitable catastrophy, and not until the human race shall have passed through it will he see the faintest promise of millennial prosperity.

The True Chronology.

But to return to our own peculiar theme. The results at which we have been permitted to arrive were only rendered possible by the previous publications of the British Chronological Society, and by them in fact, was the original idea of attempting this problem begotten. We therefore wish to testify in the behalf of their inestimable researches, and to urge upon all faithful Christians to possess themselves at once of whatever they can reach of their results. Not only, however, have we used these works very freely in our calculations, but we are particularly indebted to them for the basis of our chronological tables at the end of this volume. Most of the original data we have verified, and whatever we have added has but demonstrated the exactness of their system. We therefore endorse these publications with as little reserve as may be due to human labors for the truth, and we assert that no arguments based upon any former hap-hazard systems of chronology can be held to militate against our own discoveries, unless at the same time they overthrow the system referred to, and which has now for the past ten years, been fully set forth in their annual almanac entitled "All Past Time."

When that system is shaken, the solar system must also be moved out of its appointed harmony, and until it is so moved, Joshua's "Long Day," and Hezekiah's lengthened shadow of the sun, which have at last been identified, will henceforth be "one day known unto the Lord" and revealed unto his people for a central and perpetual *point d'appui* in Chronological Astronomy.

The writer has verified this "line of time" at all the crucial dates of history, and to the very last elements of the cycles, they work out without error in their progress, to our present day.

Beyond this testimony, and the few brief and simple arguments hereinbefore given, we cannot go at this present without introducing the unwieldy volume of abstruse and confusing figures incident upon the direct calculation. But there are a few, collateral to these, to which we must call attention ere we close.

THE SIX THOUSAND YEARS OF CREATION.

Joshua's Long Day, of 47⅓rd hours duration, from its Tuesday to Thursday's sunsets, was the last day in broad prophetical chronology which is to be wholly counted as Solar Time. That is, this particular "Day" marked an important era in the world's scriptural history, which is now to be revealed, to-wit:

Since *that* Day the millenaries have been "*shortened*" to lunar years, so that there will extend from thence $3444¾+$ lunar years to March 1899, A. D. The sum of the $2555¼+$ "long," or solar years, up to that day, and the $3444¾+^{ths}$ "shortened" or lunar years, from thence to the specified equinox, is exactly 6,000: Thus some particular day near the vernal equinox of the year 1899, A. D., will accurately terminate the 6th millenary since creation.

When it is borne in mind that the consensus of the faithful, in all generations, has anticipated that such a date will be frought with stupendous changes in the Divine method upon earth, its possible import must become apparent to all who are additionally impressed with the startling character of the days in which we live.

Joshua—Christ—Columbus.

Exactly 1441 solar years forward from that winter solstice, which thus marked one of the two most notable exhibitions of Jehovah's power which the physical universe ever experienced, brings us to the winter solstice which equally marked the culminating day in the spiritual affairs of the human race, to-wit: the birth of its Saviour; while 1441 solar years later lands us at the birth of Columbus, destined to bear the story (Christoferus) of the Hebrews to another world.

THE KEY TO CHRONOLOGY. 77

If we go backward from this Long Day for an equal period, we reach a day in Noah's life which, were not the records swept away, the mind of faith must certainly rest satisfied was quite as pregnant with import to a world then rushing onward to destruction.

All the Mosaic, and Noachic Chronology verifies itself—for they are one—upon the "line of time" which we have found to be so fully endorsed by Astronomy, to the very limit of accuracy, and the whole sequence cries aloud against those, be they fools or knaves, who would belittle them by measurements against the stature of their own littleness.

THE KEY TO CHRONOLOGY.

Another remarkable fact connected with this 2555th "year of the world," is, that it affords the key to the entire Hebrew soli-lunar Calendric system. They originally counted 7 lunar-years of 354 days each, equal to 2478 days, and then waited, or were "silent," *i. e.*, *intercalated*, 77, or eleven full weeks of days, in order to "float" the lunar on to solar time. Thus, $2478+77=2555$ days=7 Solar years, *i. e.*, $7\times365=2555$ days.

This was the fundamental "Cycle" or the most ancient Calendar, and squares itself against every date of the Bible down to the birthday of Heber. They could not have intercalated less than this number of days without severing the sequence of the week, which they did not dare to do. Nor could they have intercalated more without supererogating their Almanac, whose sole object was to keep the "generations" accurately, but at the same time in harmony with the week, the lunation, and the solar year.

But in the course of time, which culminated around the Diluvian era certain considerations [too transcendental to be referred to here, but which are fully set forth in "All Past Time" the organ of the British Chronological Association], led to the abandonment of this simple system for a cycle of "15 years" somewhat similar to our own of XIX years, and to our solar cycle of 28 years, but worked of course upon a lunar basis, and down to the very end of Hebrew history, which was swept into temporary chaos at the destruction of Jerusalem, the two systems run harmoniously through every date enumerated in the Bible.

By that time, however, other systems, notably the Roman were in vogue, with common points enough, tangential to the two, to allow of the accurate trace of time back to its origin, and it is through this now rectified line that History can confidently walk, accompanied by an accurate Astronomy and by a fearless science of Chronology.

The Unbroken "Week."

The fundamental fact which thus results is that in spite of all our dickerings with the Calendar, it is patent that the human race has never lost the Septenary sequence of the *Week days*, and that the Sabbath of these latter times comes down to us from Adam, through the Flood, past Joshua's Long Day, by the Dial of Ahaz, and out of the Sepulchre of the Saviour, without a single lapse!

No day is missing; no cycle calls for less; all call for the same, and all unite in a concert of testimony not to be shaken by any ingenuity of man, or devil.

Indeed, while with human perversity we have deliberately broken into seven pieces that primeval commandment whereby God sanctified the *seventh* day, (Gen. ii. 2), and blessed it as the chief among the seven, our very sin has conspired to keep the *sequence* of the week-days with a degree of accuracy not at all to be doubted.

Dating from Babel's confusion, men have preferred to elect their own sacred day, and down to the present time some Region, Race, or Religion has peculiarly charged itself with preserving the accurate sequence of its own peculiar day. Thus, the Assyrians kept Wednesday, the Persians Tuesday, the Egyptians Thursday, the Jews Saturday, the Greeks Monday, the Turks Friday and the Christians Sunday. The HUMAN RACE has thus kept the WEEK and has kept it intact from the dawn of time. No chronological fact is so sure as this, and in it the certainty of God's overruling power is made plainly manifest. It is but another instance of the *irony* of "Kismet."

Now, bearing the above and most ancient calendric method in mind, it is not a little remarkable that this particular conjunction, which marks Joshua's Long Day, and stands at the dividing of the Chronological scale, occurs at the first winter solstice of the 365th Sabbatic cycle of Human History, to wit: at that which followed the 365th Sabbatic year itself! That is, 365 7th years, of the world's duration, lands us at the *end* of 2554 A. M., which was the 2555th astronomical year complete. Three months forward brings us to the winter solstice of 2555 A. M. (or of astronomical duration the 2556th).

It was therefore during the opening years of such an auspicious cycle, and one so intimately related to their own Calendar, upon the scale of a year to a day, that Israel received the *guarantee* of the Land of Promise, which is yet, according to the covenant, to be made perpetual.

The End of the Age.

And now it must be briefly stated, as an inevitable concomitant of this sequence of Astronomic events, tied to Chronology and History by bonds which may not be sundered, that the 6000th soli-lunar year above mentioned, to wit (1899 A. D.) coincides with the 2520th full Solar year since Nabopolassar shook off the yoke of Assyria, and, by thus assuming the crown of Babylon, commenced the "Times of the Gentiles."*

---

* See Appendix E.

His accession took place in the 7th Civil (1st Sacred) month of the year 3377 A. M. The "Times of the Gentiles" therefore run out 2520 years thereafter, or in March 5897 A. M. (our A. D. 1899).

This opening year of the Chaldee-Babylonian Era, 3377 A. M., was "Josiah's 13th year," and was marked by the significant *"call" of Jeremiah* as a "Prophet to the Nations" (Jer. i), a fact which, in spite of the author's views recently published in the First "Study of Our Race" (The Romance of History) struck the writer almost dumb with astonishment when subsequently it was discovered.*

Moreover, the 2513th year of this Babylonian era corresponds to the 3377th year from the Exodus, and a half-year onward upon each, to-wit: to 2513½ and 3377½ respectively, repeat exactly, the famous A. M. dates of the Exodus, and accession of Nabopolassar! †

In modern A. D. style, (which, owing to the changes introduced by Pagan and Papal Rome, and by Parliament, is 1¾ years ahead in its enumeration of "Past Time"), this date corressponds to the autumnal equinox of 1892, while the seven final years of the Babylonian era, (universally believed to be those of ANTI-CHRIST!) commence at the Easter or Passover of this same

---

* See Appendix E, and Part II, Chronological Appendix.
† *Vide* Tabular Statement, showing the "End of the Age," page 207.

year, (March, 1892), according to our modern reckoning.*

A Significant Year.

It is a further remarkable fact, that this 2513th year of the Babylonian era corresponds to the 5651st upon the modern *"Jewish"* scale which year commences upon September 15th (1st day after the Harvest moon) of the *current* (1890 A. D.) year (*vide* this year's almanacs).

This number 5651 cannot be written in Hebrew without suggesting the word JEHOVAH! This is explained by the fact that the Hebrews had no figures but employed their own letters in lieu thereof, and read them from right to left.

Thus $\left\{\begin{array}{c} \text{א ה ו ה} \\ \text{H V H A} \\ 5\ 6\ 5\ 1 \end{array}\right.$ suggests $\left\{\begin{array}{c} \text{י ה ו ה} \\ \text{H V H I}\ ^{10} \\ 5\ 6\ 5 \\ (5\ 6\ 6\ 0) \end{array}\right.$

The zero at the right having no more value in Hebrew arithmetic than a zero at the left in ours. The first value above given is the more accurate one chronologically, it being only an arithmetical pointer, or suggestive of the "Tetragrammaton" or "incommunincable" Hebrew and Masonic "WORD" to wit: the name of the Almighty, Je-Ho-Va-H!

Now the writer, who is a firm believer in the plenary inspiration of the Bible, according to the

---
* *Vide* Appendices D and F.

strictest definitions of those who are called "Pre-Millennial Adventists," has been fortified in this faith by astro-chronological investigations, from whose signification he cannot escape, and he does not hesitate to affirm his conviction, as a resultant from the consensus of testimony only outlined here, that the civil and sacred Hebrew years 5651–2, dating from September of *this current year* (1890) and extending, by their overlap, to March 1892, will mark an era of astounding moment, not only to "Jews" the world over, but to such Christians as are "awake" and accept "Moses and the Prophets" literally and without the leaven of the so-called "Higher Criticism."*

A Solemn Warning.

But aside from all acceptance or non-acceptance of prophecy, these dates are OMINOUS to the whole Human Race, and they portend events for which we are as unprepared as we are to stand before the "Judge of men"—the date of whose literal ADVENT is trembling in its chronological concealment, and it almost seems certain must announce itself—although only *by its own* FACT, before this final week of years has reached its midday and meridian of terror.

It will probably be said that the writer has gone mad, and that his figures are mere coincidences, but he stands upon too firm a basis,

---

* *Vide*, Appendices G and H.

founded upon the cycles themselves, and is too deeply concerned for the Race of which he is so powerless a unit, to care a straw what some may say, if so be, by any adequate and honest means, he can persuade as many as will heed, to look unto "the Rock whence they are hewn," and to set their houses and their lamps IN ORDER.

It is to this end alone, that, guaranteed by the accuracy of the calculations which he now announces, he presumes to lift his voice in such unwelcome news to the majority of men, and he would be derelict in every duty which he owes to honesty of purpose, weighed against the magnitude of its necessity, as viewed from his own convictions, did he resist this impulse to utter a warning which in his heart he does believe is true.

And this warning is to JUDAH *in particular*— in that the events with which the days now close ahead of us are pregnant, are not to be *confused* with the Grand return unto their land, long promised in their prophets.

The compact of the *immediate* future is to be made by "many" only, and with one who is to come in his *own* name!—and woe to all who make it, (Dan. x. xi. xii).

It is a plain fact, to those who are still "wise" in their knowledge of the prophecies, that "Judah" cannot go home to Palestine with any hope of security and blessing, except she goes *in company with and under the protection of the nation* of "ISRAEL," (Ezek. xxx.vii) and in the faith and spirit of Isaiah xxvi and xliii.

A compact formed with any other people, or ruler upon earth, save one clearly identified as DAVID'S *literal* representative, can only be *in vain* and must lead to a disaster unparalleled even in their own history.

Nevertheless, into the temptation of just such a compact the trend of modern events is inevitably moving; and in spite of any warnings whatsoever, it will be made and *paid for* to the very last jot required by prophecy.

Now the circumstances which concatenate towards such an event are arranging themselves in such an apparently natural order as to promise to deceive almost the very elect. To take a single instance. There is at present no apparent motion among the Jews, looking towards any sudden awakening of a long pent up and sometime latent spirit of irredentalism.* Yet none the less, just beneath the surface every element exists, ready to spring into activity, and become world wide in its influences and results, and never were those among this scattered people in a better state of readiness for this movement. A very spark will light the conflagration.

JEWISH IRREDENTALISM.

Thus, in connection with the coming, or at least proposed Columbian Celebration of 1892 A. D., it

---
* See Appendix I.

is a significant fact that its peculiar import to the Jews, as a down trodden race seems hitherto to have escaped all general notice.

In 1492 A. D. the Jews were banished from Spain; later, from Portugal and France, and the event was considered by them to be quite as great a calamity as the Roman destruction of Jerusalem.

Taking this into consideration, together with the pointed query whether Columbus himself was not a Jew, *i. e.*, of Jewish parentage and extraction, it is manifest that all the elements exist which are requisite to make the coming anniversary a most momentous one in the opening annals of Hebrew Irredentalism.

As lately noticed in the "*Jewish World,*" no people figure so prominently in the history of the discovery of America as the Jews. The plans and calculations for Columbus' expedition were largely the work of two Hebrew astronomers and mathematicians. Two Jews, also, were employed as interpreters by Columbus, and one of them, Luis de Torres, was the very first European to set foot in the New World! When Columbus sighted the Island of San Salvador he imagined that he was approaching a portion of the East Asiatic coast, and he sent Torres—who was engaged for his knowledge of Arabic—ashore to make inquiries of the natives.

It was probably this Torres who was the Madrid Jew to whom Columbus bequeathed half a mark of silver in his will.

Another curious fact is, that it has been seriously suggested, by Dr. Delitsch, we believe, that Columbus was himself a Jew, or rather of Jewish birth.

The name Christopher was frequently adopted by converts, while the surname Colon was borne by a distinguished family of Jewish scholars.

Christopher's brother, Diego, originally bore the name of Jacob, which sounds surprisingly like a *Shem Kadosh*.

Perhaps, during the preliminaries to the coming celebrations, some Jewish scholars in Italy will make inquiry into the validity of this daring suggestion, and at any rate the vast import of the coming years to Judah, as at last almost a liberated race, cannot be gainsaid, nor is it at all extravagant to think these years will witness a final effort to complete and celebrate their full emancipation.

The fact is we have already entered upon a decade filled with Jewish centennials, the import of which cannot but increase as they return. To refer for instance to but a single case: 100 years ago this year, the National Constituent Assembly was formed in Paris, and one of their earliest acts (1790) was to declare the Jews of Spain and Portugal to be Citizens of France. Thus, for 300 years, the Sephardim had been without citizenship in those countries, when the nation, whose Napoleon but a few years later reconvened their Grand Sanhedrim, helped them to celebrate their tri-centennial by an act of emancipation, whose own centennial we are now calling to mind! Let him, therefore, who sees nothing significant for "Judah," in the years now coming into general history, cast up his history by cycles, and by centennials, and, if he be at all a "Jew," he will find sufficient to amaze him. Indeed, if we read aright the still latent portents, it is to this very land of France, and to a shadowy Napoleon, that they still significantly point—and yet, withal, with OMINOUS significance!

THE LAST KING OF THE FRANKS.

In this connection the following words of St. Augustin, written *circa*, 400 A. D., are of peculiar significance, to-wit:

"Quidam verò doctores nostri dicunt, quòd unus ex regibus Francorum Romanum Imperium ex integro tenebit, qui in novissimo tempore erit. et ipse erit maximus et omnium regum ultimus : qui postquam regnum suum feliciter gubernaverit, ad ultimum Ierosolymam veniet, et in monte Oliveti, sceptrum et coronam suam deponet ; statimque, secumdun sententiam prædictam apostoli Pauli, Antichristum dicunt adfuturum."— *Op. Divi Augustini, ed. Paris.* 1685, *t. ri. p.* 244.

Which being interpreted is as follows:

("Certain of our scholars assert, that a king of the Franks will possess the Roman Empire *restored*; which king will come in the last time, and will himself be the last and greatest of all kings ; who, after having ruled with success, at the last shall go to Jerusalem, and shall lay down his sceptre and his crown on the Mount of Olives, and they add that immediately, according to the above cited prediction of the Apostle Paul, Antichrist will appear.")

We need not refer Bible students to the numerous commentaries which fill the library of prophetic exegesis, wherein it is harmoniously agreed that out of France, in her mysterious role as the perpetuator of the Roman empire, APOLLYON, the Beast of Revelations, is yet to re-arise, and who, healed of his "deadly wound," (Rev. xiii) is still to consummate his part in human history—the chapters of that history, now about to be revealed by Facts, reserve the privilege of being their own interpreters.

THE CONTROVERSY OF ZION.

But the most potential element in Judah's latent possibilities is that which underlies the final solution of the "Eastern Question"—to her it is purely the "Controversy of Zion;" and, no matter how it may be viewed by the rest of the world, it is impossible that Palestine shall be emptied of the "unspeakable Turk," and the fact not create a furore among the seed of Abraham.

That all things have been ripening towards the speedy settlement of that question is patent to statesmen, and that Russian Statesmen, from their own purely *ex-parte* interests, are bent on hastening its solution, is the most evident fact now upon the political horizon of Europe.

It is the very fact that no one can tell when this ever agitated topic shall be sprung for final settlement, which makes its resultant possibilites so momentous, and from now on we may scrutinize the Bulletins for events *whose sequence human wisdom cannot fathom* WITHOUT REFERENCE TO THE SCRIPTURES. In other publications, (*vide* "Yale Military Lectures" 1890, the discussion of "the Eastern Question" in *Frank Leslie's Weekly*, of April 12, 1890, and in a complete series thereon, published in the New Haven *Register* of March and April, 1890), we have fairly covered this topic. But the true philosophy of the situation has been set forth best in Study No. 1 of this Series—the "Romance of History"—now within the reach of any who are interested in obtaining information upon the broad issues involved.

A MIDNIGHT CRY.

In 1837 the Kingdom of Heaven was likened unto the ten virgins, five of whom were wise and five foolish, who took their lamps, and, in 1844, went out to meet the bridegroom, and like them, because he tarried, lo, we fell asleep!

But it is time to wake!

Reckoning from 3466 A. M., when Daniel uttered his remarkable prayer (chapter ix), the gloom upon the Dial is close upon its MIDNIGHT MARK—so dense the darkness that it "may be felt."

Sleepers, awake!

There is barely time to trim your lamps!

The long expected "midnight cry" is breaking on the ear!

"𝔅𝔢𝔥𝔬𝔩𝔡 𝔱𝔥𝔢 𝔅𝔯𝔦𝔡𝔢𝔤𝔯𝔬𝔬𝔪 𝔠𝔬𝔪𝔢𝔱𝔥!"

𝔊𝔬 𝔶𝔢 𝔬𝔲𝔱 𝔱𝔬 𝔪𝔢𝔢𝔱 𝔥𝔦𝔪!

"*Ye do ERR,*

Not knowing the Scriptures,

Nor the POWER of God!—Matt. xxii. 29

PART II.

APPENDICES.

"*Precept upon Precept; line upon line, line upon line; here a little, there a little.*—Isaiah. **xxvi.**10.

APPENDIX A.

THE BOOK OF JASHER.

I quote the following from Smith's Bible Dictionary: " Jasher, Book of, or, as the margin of the A. V. gives it, 'the book of the Upright,' a record alluded to in two passages only of the O. T. (Josh. x.13 and 2 Sam. i.18), and consequently the subject of much dispute. That it was written in verse only may reasonably be inferred from the only specimens extant, which exhibit unmistakable signs of metrical rhythm. Gesenius conjectured that it was an anthology of ancient songs which acquired its name, 'the book of the just, or upright,' from being written in praise of upright men." Thus far the Rev. Dr. Smith :

Of course this book has been a subject of dispute, everything in the Bible has been, but never should have been *within the fold.* The whole tenor of this article, like so much else, written both within and without the fold, is misleading and unwarranted. What should we care for the mere conjectures of Pharisees in matters of faith. Take for instance that of the doctrine "that it was written in verse *only.*" To show the common sense man how unwarranted his reasons are for considering it all in poetry, let us refer to Genesis—purely a book of Hebrew prose, yet containing two distinct pieces of poetry (chap.

iv, 23, 24, and chap. xlix). Now suppose that Genesis had been lost, like Deuteronomy was, and yet that other books had contained these two poems, the case would have been similar to the loss of "Jasher," and our "doctors" would have pronounced it a mere poem, and so have lent color to the teachings of the School of Disbelief. But, again, and on the basis of commonsense, what if it were in poetry? So too is more than one-third of the Bible, *i. e.*, the bulk of the prophecies from Job to Malachi inclusive! I would to God that all that has ever been written within the church, that has savored of apology, or been tainted with concession to the outside, had shriveled into SIXES as it dried, that men might know now where they stand, and what their teachings really imply. As to the book of "Jasher" they know nothing save what "is written," and it is their duty to say so plainly, or at least to maintain strictly only what is "sound" lest it betray some weak one into deadly error. Finally there is nothing in the reference to "Jasher" in Joshua x to imply that his command v. 12, is even a quotation from the book of Jasher. The question, "Is this not written in the book of Jasher?" (v. 13) is a reference to the book, *then in existence*, by direct implication of the context; or, at least, until the book is forthcoming our own explanation is as good as the collateral, and we warrant will suit the commonality of men far better.

APPENDIX B.

Casual Eclipses.

We are perfectly aware that "casuals" in the eclipse line, may alter this number, and make it more or less. We did not have a smoked glass to our eyes when "Father Time" went through his phases—what we mean is, that since the dawn of the Mosaic era this number of normal eclipses fills the measure, nor can it be altered until that era has lasted long enough to have pointed out to us by experience the grander law of "casuals," whose equation, probably, has not yet integrated once.

APPENDIX C.

The Earliest and the Latest Eclipse.

A few points respecting this *last* eclipse of history (Tuesday, June 17, 1890) are here in order. It was an annular eclipse of the Sun and was not visible on this (American) continent, save at the most eastern coast of Brazil. Its central line commenced on the Atlantic ocean, in Lat. 5° 8' N, and Long. 32° 30' W. Passing easterly it struck Africa near cape Verde, thence crossed the Great Desert, struck the Mediterranean near cape Bon, crossed Asia, passing over Turkey, Persia, Afghanistan, Hindostan, and terminated in Indo-China. 2,604 years ago this same eclipse must have pursued the same identical path, and been visible to Sargon at the Siege of Samaria! It is No. 45 in the regular team, and occurs every twelfth year of the eclipse cycle, but only repeats its last modern dates at intervals of 651 years. That it did thus repeat its modern dates in 3284 A. M., when Sargon might have seen it, is absolutely demonstrated by the fact that Ptolemy records two succeeding total lunar eclipses (No. 48 and 50) at Alexandria, and refers to them as *repeating* the same eclipses seen at Babylon in 3284-5—the whole sequence is thus proved. It is on evidence such as this that students of the *true* chronology, which is the *"Biblical* chronology," take their stand, and are unconcerned at the sneers, and inuendoes of any criticism which is founded upon a Science bound to pass away since it is merely masquerading nowadays upon false premises. Finally, as to this eclipse when in its normal route (as at its last occurrence, and in 3284 A. M.), it was absolutely central as to time and terrestrial locality, upon the Prime Meridian of the Great Pyramid ($30°+$ E. of G.), at 21^m 58.5^s after Greenwich mean noon of Monday, June 16th, in north latitude 36° 40.4' or in the zenith of Lycia in Asia Minor.

The interesting circumstances of the *first* eclipse of history (No. 1, Solar) of the regular team, have been calculated by the Premier Chronologist of the British Association (Mr. J. B. Dimbleby, The Shrubbery, Chatham Place.

South Hackney, Eng.) It is found to have occurred upon Friday, the 1st day of the 4th civil month of the year 0 A. M. Its repetitions were recorded by the ancients in 3401 and 3419 A. M. (Judah's date of Captivity being 3406 A. M., *i. e.*, between the two consecutive records). Both of these eclipses occurred on Friday, upon which week-day this eclipse still repeats every 651 years, *vide* its re-occurrence upon Friday, Jan. 11, 1861, *sunset* reckoning from "*primeval meridian.*" This latter repetition was 9 × 651 years from the Adamic original.

Thus: date of 1st eclipse Friday, 1st day, 4th mo., *i. e.*, our January). 0¼ A. M.

Add 651 × 9 = 5859 (years of duration).

Last repetition 5859¼ A. M.

Reduction to 1¾

A. D.   5861   = 1861 A. D.,

Jan. 11th, Friday.

This alone is a demonstration that the sequence of the week-days has never had a lapse, and is but a single one of many collateral "lines of time" which run back harmoniously to the Biblical chronological starting point, as unavoidably a Sunday, the origin of the Solar Cycle, of the common team of eclipses, of the lunations, of the equinoxes, of the diurnal cycle, and all their transcendental combinations. It is thus clear, (whatever views we may individually hold as to the creation, or re-fabrication of the earth), that this Mosaic starting point is a *scientific* one of the most unique order, a common cusp in all the exponential equations of astronomy, and of chronology which is its practical counterpart, so well as the literal origin of Scriptural History. Human existence is rigidly tied, and circumscribed between its two extremes, the ancient one a point beyond which all is chaos, the modern one always evolving new combinations which individually and in concert rigidly reverse to the one and only origin.

It is on this account that we have adopted the A. M. (Anno Mundi) years as the truly scientific ones.

Finally, as an indication of the supreme importance of the eclipses to chronologists, we may quote the following:

"Such is the precision of the periods of the eclipses, and the continued accuracy of the length of the day, that a particular eclipse (solar) is always visible from the same part of the earth. The ancient records of Eclipses in Nineveh are now total in that part of Asia where Nineveh stood, the track of totality is precisely what it was nearly 3000 years ago."—J. B. Dimbley, Premier Chronologist B. C. A.

APPENDIX D.

Changes in the Times and Seasons.

The Romans changed the beginning of the year from September to March, and the Calendric year which preceded this change was therefore but 6 months long, *i. e.*, the then current A. M. year had run from September to March, when the New Era began as 1st year of Rome. In the same way when Parliament changed the beginning from March to January the year of change was but 9 months long, *i. e.*, the current year (1752 A. D.) had run from March to January (9 months) when the new era began to count as 1753 A. D. Thus the Romans got ahead of A. M. time, in so far as their enumeration of its years are concerned, by 6 months, and Parliament by 3 months more, making a total of 9 months ahead.

But there was a graver error introduced into the count by the Abbot Dionysius Exiguus who first instituted the system of dating the Calendar of Time from what he supposed to be the birth of the Saviour. This was done in 527 O. S., 4525-6 A. M., and in the system thence resulting we call 4000 A. M. (instead of 4001) our A. D. 1, thus putting ourselves a whole year more *before* true time. The sum total of these three "changes of the times and seasons" places our count, in round year numbers, 1¾ years ahead of the true count. Hence to correct this error we must always deduct 1¾ years from any A. D. date. Thus, the Autumnal Equinox of Sept. 22, 1889 would equal, if we were right, 5889¾, but as we are 1¾ ahead it equals 5888 A. M., and brings us to the *original* "New Year's day," and it is to be noted that, as A. M. years denote *past* time, this new year's day ends 5888 years then and there scored off, and is the *beginning* of 5889 A. M., or the 5890th year of astronomic duration since creation — the year which terminates with the date of this publication.

On account of this confusion, which we ourselves, in our ancestors, have introduced into the Calendar, it is absolutely impossible to tabulate the Eclipses and other cycles in a consecutive system of modern A. D. and B. C. years. The cycles laugh them to scorn.

But it is far different when we work them upon the Scientific A. M. line! They then obey,

"Day unto day uttereth speech,
And night unto night sheweth knowledge.
There is no speech nor language where their voice is not heard.
Their line is gone out through all the earth
And their words *to the end of the world*.
In them hath he set a tabernacle for the sun
Which is as a bridegroom coming out of his chamber,
And rejoiceth as a strong man to run a race.
His going forth is from the end of heaven,
And his circuit unto the ends of it;
And there is nothing hid from the heat thereof."
                                            Psalm xix.

It is thus manifest how greatly those err, who, basing their natural deductions upon the confu-

CHRONOLOGICAL CONFUSION. 103

sion of Chronology as commonly understood, declare that "time" is arbitrary, and that history, particularly Sacred history, is inextricably confused when measured against the Cycles. This is true when the chart that guides our ship is stamped with the *visé* of our human legislation. No sooner have we passed along than our wake is lost amid a chopping sea. But when we cast these Jonahs overboard and move a point and three-quarters to the right, the eternal Cynosure shines on our course and is reflected forever in the phosphorescent pathway which we leave behind.

Thus the Scientists and Legislators have actually introduced an error of $1\frac{3}{4}$ years into the Christian Era now employed.

But there is a practical side to the effect of this error upon the common mind, which acts in ignorance of it, that is worthy of a special showing, since it discloses a curious irony of Providence, whereby the ordinary mind is enabled most accurately to arrive at the true year of the nativity.

The human race has never varied its *method* of keeping account of its own "age." For instance, a child born upon a 25th of December is not counted "one year old" until it attains unto its next birthday, and thereafter, until it is at still another birthday it remains "one year old." Thus we say "he is one year old," or "*in* his second year," until he reaches his third. At death, however, the final year, *i. e.*, the current one up to the moment of demise is always taken into account. A person, for instance, is "39 years old," he is therefore "in his 40th year," and in fact, at the date of this publication will be 39 years 7 months and 19 days. Hence, according to the common reckoning he was born in 1851, which is close enough for our illustration of the count by years.

Let us now (ignore the 1¾ years error in the current era, and) consider what we call the "years of our Lord" to represent the Saviour's AGE. On the supposition of accuracy the common reasoning is as follows:

"It is now autumnal equinox; the year is therefore 9 months old; if the Saviour were still alive, *i. e.*, if his *earthly* life had not been broken, he would be 1890¾ years 'old,' therefore, he was born 1891¾ years 'ago;' the Christian era therefore commences 0, 1, 2, 3, etc., and runs down to this time, which, so far as *duration* is concerned, is 1891¾."

Now such a method of reasoning, although fallacious, would land the common intellect at the right A. M. year, to-wit: 3996 A. M. and would actually run parallel to the correct A. D. years, and synchronize with them all along the line. Or his calculations might be even ruder, and as follows:

"They say that 5889 A. M. ends to-day: Then our year ought to end with it, and be 1891 instead of 1890¾. If so, the Saviour would be "1891 years old," or at his 1892d year. Therefore subtracting 1892 from 5889 leaves 3997, which must be his year of birth."

Or finally, if he simply subtracted 1892 from 5889 he would obtain 3997 which in reality *is* the year of "astronomic duration" that corresponds to 3996 A. M. This is patent from the fact that A. M. years consider "*past* time" only. Hence, when the world was "3996 years old" it was "in" its 3997th year.

But all of this unfortunate confusion is easily avoided by discarding at once every system except that founded on the natural and original "years of the world" (A. M.). Upon this system astronomy "works" forward and backward without hitch, and corroborates sacred and secular history. Nor will the three work together upon any other system.

APPENDIX E.

The Biblical Cycles all Astronomical.

It is fitting at this juncture to call attention to some of the beauties hidden in the years and periods familiarly employed by the prophets, and which the generality of men and some ministers account as not only of no significance, but even more, as mere inventions whose chief object is to awe the timorous into subjection to the Hierarchy.

There is not a period mentioned in the Prophets which is not an astronomic cycle of consummate use. Let us take but one, which is the source of many of the rest. A "Time" is 360 years, and it is employed consistently by the Spirit in its predictions as to human affairs. Moses, Daniel, John, all couch momentous prophecies in terms of it, and it is most generally known to us in its maximum value "Seven Times," or a *week* of "Times," *i. e.*, 7×360=2520.

Now the first notable point with reference to this number is that it is the "least common multiple" of the decimal system, *i. e.*, it is the smallest number which is divisible without a remainder by each of the digits. This alone shows that there was no accident in its selection.

2d. It is exactly 140 eclipse cycles 18 years each.

3d. It is 168 Ancient Hebrew solar cycles of 15 years each.

4th. It is 360 antediluvian solar and Sunday cycles of 7 years each.

5th. It is exactly 90 modern solar cycles of 28 years each.

6th. It contains 132 Lunar or Metonic cycles, in which the "epact" amounts to 77 *lunar* years: and over and above these cycles there is a remainder of 12 years, which raises the "epact" to just 75 solar years. "Now here we are confronted with another startling fact, a fact which it will puzzle the ingenuity of skeptics to account for," *and* a fact of vast astronomic import.

THE BIBLICAL CYCLES. 107

"In the last chapter of Daniel the Angel intimates to the prophet in answer to his chronological inquiries, that while the scattering of the power of the holy people should terminate at the end of the second half of the 2520 years, yet there should be additions of 30 and 45 years before the era of full blessedness would arrive (Dan. xii, 11–13). In other words, to the long period of 2520 years, Scripture adds a brief period of 75 years, and as we have just seen, astronomy does the same. The difference between 2520 true lunar and the same number of true solar years is seventy-five years. In other words, the 75 years added in prophecy is exactly equal to the "epact" of the whole "seven times." But to exhaust this subject would be to write an encyclopedia. The years of Daniel, like the Creation, the Exodus, the birth of Christ, etc., was the commencement of a common team of eclipses, his "time" is 20 cycles, and his "time, times and half a time" (1260 years, or half of "7 times," i. e., half of 2520 years) is 70 cycles of 18 years, the period when a common team comes around. To mention but one other fact: 315 years, (which is a quarter of 1260, and is employed upon the great scale as a measure of each of the 8 working hours of prophecy), is itself a soli-lunar cycle ten times more accurate than the Metonic cycle. 1260 years is itself such a cycle, after which the sun and moon return within less than half a degree to precisely the same point of the ecliptic, and that within an hour of each other. That is, it is a soli-lunar *diurnal* cycle; and so is Daniel's 2300 years, and affected with the same slight error. Their difference, 1040 years, is such a cycle with an error of but 1 hour!

Not the least remarkable cycle hidden in the Scriptures is the one concealed in the date of the Saviour's birth. This has now been fixed beyond all dispute as falling at the winter solstice of 3996 A.M. It now remains to point out a *fact* which the writer's own studies have revealed. 3996 = 6 × 666! That is, the birth of Jesus Christ, in whom we are saved, occurred at the very *crisis* of man's "death" inherited in Eden. We cannot begin to summarise, even briefly, the host of tangencies, all along the stream of the true chronology, which come out from the use of this period as a divisor of the years of our disease, and of our regeneration. This factor of the A. M. years is literally a "day" upon a cycle which plainly records the moral sickness unto death, of the Human race, in exactly parallel terms to those employed by physicians in discussing the septenary progress of bodily ailments, all of which run in parts or multiples of weeks.

So too, though not now relatively remarkable, since all about the true chronology is an astonishment, the Saviour's birth occurred at the 1st year of the 222d eclipse cycle. In other words, 6 greater cycles (6 × 651 years) were past, and the 1st year of the 6th shorter cycle was then current. No wonder, therefore, that the Holy Spirit has seen fit to characterize the seal of man's deadly adversary as **666** (Rev. xiii.), and has warned all men against accepting any privileges to "buy or sell" (17) by virtue of "a mark" (16) which shall spell and count the name of "Anti-christ" (18), under penalty thereby of sinning hopelessly (Rev. xx. 4). "Herein" verily "is wisdom," and may the strength of Him who died that we might live, be with us when we are called upon to give up what is his, rather than to live simply unto death indeed. That we are near this final crisis in the "Mystery of Iniquity" should be patent even to the common mind when we refer him to the almanac and point out the following facts: we are (Sept. 1890) in the sixth year from the end of the current solar cycle of 28 years duration. At the end of 1895, when this current cycle terminates, exactly $66\frac{2}{3}$ (= 66.66666 + etc.) such cycles of 28 years, or $1866\frac{2}{3}$ solar years, take us to the Baptism of the Saviour—$i.\ e.$, to the commencement of his ministry unto souls needing a Physician, and the studies of all who have devoted themselves to the "Signs of his Coming" agree that it may be confidentially looked for any time from now on until then.

APPENDIX F.

CHRONOLOGICAL ERAS HARMONIZED.

"The era of Anno Domini, commonly abbreviated A. D., was invented by Dionysius Exiguus about 527 A. D. [4525-6 A. M.] It was ordered to be used by the Bishops by the Council of Chelsea in 816 A. D. It was not generally used for several centuries. Charles III of Germany was the first monarch who added 'In the year of our Lord' to his reign in 879 A. D." Dict. of Dates, Hayden. All of the above years except the A. M. years in brackets, are "old style:" so is the frequent reference to "753 of Rome," or to the "30th year of Augustus Cæsar."

As already stated, we have purposely avoided all reference to the B. C. and A. D. years. There is no knot which has been so snarled as that of Chronological *duration*, and the kernel of the whole confusion centers about the A. M. year which marked the Saviour's birth. The true date of that Nativity was at the winter solstice of the year 3996 A. M., which year had a Sunday autumnal equinox; *i. e.*, it was 8 years earlier than Usher's common date (4004 A. M.), or 5 years before the 4th millenary commenced. This is proved by central solar eclipse No. 1, occurring in 3996 A. M., which was followed in due time by total lunar eclipse No. 8, Jan. 18-19th, 3998, A. M. This latter eclipse was recorded by Jo-

THE SAVIOUR'S BIRTH. 111

sephus, and "decides the period of birth, and the entire chronology of Jesus Christ. According to this Jewish Historian, Herod put a priest to death on the night of this eclipse of the moon, ('That very night there was an eclipse of the moon.' Josephus, Antiquities, xvii. Chap. V. Sec. 3), after which, he being near his own death, shut up some eminent men of Judea in the hippodrome, ordering them to be killed as soon as he died. * * He died on the 5th day after putting Antipater to death, having reigned 34 years since he procured Antigorus to be slain, and 37 years since he was declared king by the Romans. Consequently this eclipse must have taken place when our Lord was about 2 years old, as Matthew ii, 16, informs us. * * This too explains the phraseology of St. Matthew concerning the slaying of the children in Bethlehem 'two years old and under, according to the time,' * * for when this eclipse occurred our Lord was two years old according to Solar time, and under two years by Hebrew time." *Vide*, "All Past Time."

To close this part of the discussion, and furnish students a chart whereby they can hereafter translate the several systems now in vogue into the correct A. M. years, we submit the accompanying Harmony.

Without such a diagram, a pilot himself could not steer the craft of History through the reefs which now abound in Modern Libraries.

An examination of the table will show two general subdivisions, to wit: *true* and erroneous systems. Under the latter we classify the years of Rome because they commence in March instead of at autumnal equinox, whereby (as shown above, *vide* Appendix D) an error of 6 months was introduced into the enumeration of "Past Time," in the year 3246 A. M.

Upon the same principle the Julian Period is also erroneous, since it was only invented lately by Scaliger, in order to remove ambiguities in the common Anno Domini years, and because, while fully accomplishing its object, it necessarily runs *with* those years, and therefore inherits their own error of 1¾ years overplus. As a chronological scale or period, it consists of 7,980 years, and is formed by multiplying together the number of years in the solar, lunar, and indiction cycles ($28 \times 19 \times 15 \times = 7980$). It is reckoned from 4713 − 14 B. C. (common), when the first years of these several subordinate cycles of our calendar would have coincided, (*i. e.*, 713 years before Creation!) The Julian Period would thus have *begun* its year, lunation and indiction, upon a Tuesday (!) instead of a Sunday (Gen. i, 4), which Tuesday reckoning from Tuesday, June 17, 1890, was 2,411,536 days ago, or 344,505 weeks + 1 day, which latter day was the Tuesday origin specified.

Upon this same scale (J. P.) the conjunction of Joshua's Long Day was upon Wednesday the

THE BIRTH OF JESUS, "THE CHRIST."

The 26th of the 3rd Civil Month, 3996 A. M., i. e., of the IXth Sacred Month. In the "Evening" of the Sabbath Day. Upon our Friday Night, December 25th, 750 a.u.c.

THE VARIOUS SYSTEMS OF CHRONOLOGY HARMONIZED.

Trask.	THE ERRONEOUS SYSTEMS.		THE TRUE SYSTEMS.					REFERENCES AND REMARKS.
	"Common," A. D. B. C. Consult The A. V. Bible.	A. U. C. Julian 'Period.'	A. U. C. The Roman Era.	A. N. Years	B. N. Years	A. M. Years 'Past Time'	Astro. Years 'Dura- tion.'	Years by Millen aries.	
14		11-10	4703-4	743-4		6	3990	3991	3990	
13		10- 9	4-5	744-5		5	3991	2	1	
12		9- 8	5-6	745-6		4	3992	3	2	
11		8- 7	6-7	746-7		3	3993	4	3	
10		7- 6	7-8	747-8		2	3994	5	4	
9		6- 5	· 8-9	748-9		1	3995	6	5	Luke i. 6.
8		5- 4	9-0	749-0	0	0	3996	7	6	ii. 1-39. ☉1.
7		4- 3	4710-1	750-1	1		3997	8	7	Matt. ii. 1-15.
6	0	3- 2	1-2	751-2	2		3998	9	8	ii. 16. 18. ●8.
5	0- 1	2- 1	2-3	752-3	3		3999	0	999	ii. 19. 23. This, 753 a.u.c., is the 30th year of Augustus Cæsar.
4	1- 2	1- 0	3-4	753-4	4		4000	4001	1000	
3	2- 3	0	4-5	754-5	5		4001	2	1	●20.
2	3- 4		5-6	755-6	6		4002	3	2	
1	4- 5		6-7	756-7	7		4003	4	3	
0	5- 6		7-8	757-8	8		4004	5	4	
	6- 7		8-9	758-9	9		4005	6	5	
	7- 8		9-0	759-0	10		4006	7	6	
	8- 9		4720-1	760-1	11		4007	8	7	
	9-10		1-2	761-2	12		4008	9	8	Luke ii. 42-52. The Saviour twelve years old.
	10-11		2-3	762-3	13		4009	0	9	
	11-12		3-4	763-4	14		4010	4011	10	
	12-13		4-5	764-5	15		4011	2	11	
	13-14		5-6	765-6	16		4012	3	12	
	14-15		6-7	766-7	17		4013	4	13	
	15- 6		7-8	767-8	18		4014	5	14	
	16-17		8-9	768-9	19		4015	6	15	
	17-18		9-0	769-0	20		4016	7	16	
	18-19		4730-1	770-1	21		4017	8	17	
	19-20		1-2	771-2	22		4018	9	18	
	20-21		2-3	772-3	23		4019	0	19	
	21-22		3-4	773-4	24		4020	4021	20	
	22-23		4-5	774-5	25		4021	2	21	
	23-24		5-6	775-6	26		4022	3	22	
	24-25		6-7	776-7	27		4023	4	23	
	25-26		7-8	777-8	28		4024	5	24	
	26-27		8-9	778-9	29		4025	6	25	Luke iii. 1, Mark i. 11.
	27-28		9-0	779-0	30	▲	4026	7	26	Mark i. 12-13, Luke iv 1-13, iii. 38.
	28-29		4740-1	780-1	31	3½	4027	8	27	John v. 1.
	29-30		1-2	781-2	32		4028	9	28	i. 4.
	30-31		2-3	782-3	33	▼	4029	0	29	xiii. 1. Crucifixion.
	31-32		3-4	783-4	34		4030	4031	30	
	These years continue down to our 1890, A. D.	=	And these to its corresponding 6603, J. P.	&c.	35 36 37 38 39		4031 4032 4033 4034 4035	2 3 4 5 6	31 32 33 34 35	Acts ix. 23. xi. 25.

1,194,006th day, *i. e.*, 173,932 weeks + 6 days before June 17th, 1890.

The inaccuracy of the "common" B. C. and A. D. years will be apparent by consulting the references in an annotated Bible (Authorized Version). For instance, opposite to Luke ii. 1, 7, 34, in a Reference Bible, it will be found stated that the Saviour was born "the fifth year before the accepted account called Anno Domini."

Passing now across the 7th line of the Table, where this 5th year occurs under the "common" B. C. system, to the true systems, their superiority will be at once apparent. The year of the world (A. M.) was 3996, which is of course the correct one from which to calculate years *before* and *after* "the Nativity." (B. N. and A. N.) However, it cannot but be far clearer that the direct sequence of the A. M. years themselves affords us the safest and most natural skeleton upon which to string the actual events of human history, since not only do we have to translate every other scale into them, but chiefly because, as we hope to show in future publications, every date in the Bible at once yields up its secret when measured thereupon.

APPENDIX G.

NOTABLE ASTRONOMICAL EVENTS OF 1891-2 A.D.

In addition to the considerations already enumerated in this paper the years 1891-2 will be of remarkable astronomic import because of four notable events which are then due. The first will be a transit of Mercury circa May 9th, 1891, agreeing with a corresponding one which occurred in the year 1 A. M., and was repeated in the year 3 A. D. At the moment (11h. 55m. 29.3s. Greenwich mean time), of exterior contact of *ingress*, the Sun will be in the zenith of 179° 48' longitude west of Greenwich, and of 17° 36' north latitude; and at the moment (16h. 52m. 45.7s.), of exterior contact of *egress*, the Sun will be in the zenith of longitude 105° 53' east of Greenwich, and of latitude 17° 39' north. The transit over the sun's disk will be partly visible at Washington, D. C., and visible throughout the western portion of North and South America. These transits are of peculiar value to Chronology while those of Venus are more particularly related to the determination of solar distances. There will be but one other transit this century, to wit, one of Mercury upon Nov. 9-10, 1894 A. D.

The second and third astronomical events of importance in 1891 A. D. will be two total eclipses of the moon, in May (23d) and November (15th) respectively, and which are of peculiar interest because they are repetitions of the same ones (No. 48 and 50) which were seen by Ptolemy at Alexandria, and duly mentioned in his *Almagest*.

But these eclipses are of still greater import in relation to "the Times of the Gentiles," since, as Ptolemy correctly informs us, they were seen at Babylon in the years 3284-5 A. M., and, consequently, at the siege of Samaria!

Finally the fourth, and by far the most important astronomical event, will be the re-appearance of the most splendid celestial body ever recorded, the variable star in the Constellation of Cassiopea! It was seen by Loviticus in 945 A. D., again by Jean in 1264 A. D., and finally by Tycho Brahe in 1572 A. D. It is believed to reappear in alternate periods of 308 and 319 years, and consequently may be looked for in the Fall of 1891 or Spring of 1892, when in the course of a few weeks it will become brighter than the planet Jupiter.

Tycho Brahe, who was one of the most eminent astronomers of his day, describes the appearance of this star in 1572 as very sudden. He says that on returning home on the evening of Nov. 11, 1572, he was surprised to find a group of country people gazing at a star, which he was sure did not exist half an hour previously. It was then as bright as Sirius, which is the largest star in the heavens, and could be seen during the day. It continued to *increase* in brightness until it surpassed the planet Jupiter! Its brightness began to diminish in December, and so continued to diminish until in March, 1574, when it wholly disappeared. It had no sensible motion, nor any parallax, and therefore must have been far more distant than the planet Neptune. Its light, which was at first white, changed as it decreased to yellowish, then to ruddy, and finally to a livid white.

APPENDIX II.

Is the Bishop of Rome a Prophet?

Anent the cry which, in solemn and sober earnest, we are raising in this volume, we submit the following significant editorial from the *"Los Angeles Churchman"* of July, 1890:

"Apropos of the opinions which we quoted in our last issue on 'The Signs of the Times,' we clip the following from the Los Angeles *Times* of recent date·

THE POPE HAS A PRESENTIMENT OF COMING EVILS.

DUBLIN, June 13.—[By Cable and Associated Press.] The *Irish Catholic* states that the Pope in replying to the congratulations of visitors at the Vatican, expressed himself as strongly of the belief that a great punishment was impending on society for its disregard of and indifference to the church.

"The Lord," he said, "will come no longer with a sweet, peaceful face, but with an angry one to strike and purify His church. I am neither a prophet nor the son of a prophet, but I feel in my heart a sorrowful presentiment. A sea of evil is about to beat against the rock on which the church is founded, and will leave nothing to be seen on the horizon but the threat of the anger of God. Prayer will not suffice to appease the Almighty."

"The profane may say the opinion of the Pope is nothing more than the fretful complaint of a disappointed old man because the world no longer does homage to the Holy See. Perhaps it is no more, but one opinion is as good as another, and others, not profane, who reverence sacred offices and functions, may recall the sayings of one who sat in the High Priest's office while Jesus of Nazareth lived. 'Ye know nothing at all, nor consider that it is expedient for us, that one man should die for the people, and that the whole nation perish not.'

"Then the divine record goes on to say: 'This spake he not of himself, but being High Priest that year, he prophesied that Jesus should die for that nation, and not for that nation only, but that he should gather together in one the children of God that were scattered abroad.'

"If God recognized the office which Caiaphas held, and employed him as a true prophet, He may surely recognize that which Leo XIII holds, who is a much better man than Caiaphas was. Whatever opinion we may hold of the claims of the papacy, the fact is undeniable that he is the lawful Bishop of Rome, an apostolic diocese, and a city which for 2000 years or more held the world's fortunes in the hollow of its hand.

"Leo speaks from his heart, as Caiaphas did, but in a much kinder spirit, and whether he does not, as Caiaphas did, echo the voice of God, is for the future to determine. Certain it is that no man living to-day exercises a wider influence throughout the world than the Bishop of Rome, whatever they may think of his infallibility, and since his adherents have tied themselves up to that opinion they may be held responsible for these heart-breathing presentiments of an aged man, who confesses that he speaks not *ex-cathedra*, but from that inner consciousness where the Lord alone sits in the temple of the heart.

"If this report, as cabled, be true, it is very significant."

APPENDIX I.

Judah Homeward Bound.

POSTSCRIPT.—It should be manifest, from the character of the tables and calculations in this volume, that they must have been the result of years of close application, and that our conclusions could not have been evolved during the past few weeks which covered their publication. At the time, early in June, when this manuscript was put into permanent shape, the remark made in the body of the work that "There is at present no apparent motion among the Jews looking towards any sudden awakening of a long pent up and sometime latent spirit of irredentalism" (page 85) was literally true. But now (August 1890) even before we have completed our proofreading of the pages which all along we have intentionally designed to synchronize with the Autumnal Equinox of 1890, a few weeks hence—the whole aspect of the Jewish situation has suddenly changed, or rather has given positive promise of the certainly coming change! We have scarcely time to sketch the altered outlook in a brief and final appendix. In the midst of press work we received the following letter, and its perusal will recall to others the suddenness with which the threatened enforcement of the Russian Anti-Jewish Edict of 1882,—the which had slumbered so long as to become almost forgotten—has burst upon the world.

NEW HAVEN, CONN. August 8th, 1890.

LIEUT. C. A. L. TOTTEN—

Dear Sir: In this same mail I send you a copy of the "National Tribune," containing an article on the coming exodus of the Jews from Russia, which I thought would be of interest to you in view of the fact that, when I called on you in last June, you told me to watch for movements in the Jewish world that would take place in a short time. This article shows a very remarkable fulfillment of your prediction, the truth of which I trust will be even more fully vindicated in facts which will come to light in the near future.—Very truly yours,

69 Lake Place.   C. C. COLEMAN.

[At this point it is but just unto ourselves, and to the earnest school of Bible students to which we belong, that we should candidly disavow any right or pretension to the "role" which authorizes one to "predict" as such, and in the sense commonly understood by the world. This is not our position. In common parlance, we are simply *believers in the predictions of those who were by Divine commission the Prophets of* "OUR RACE," and the utmost that we dare to assume unto ourselves in these premises, is a clearer understanding of these prophecies than certainly seems to be the common property. The key to this broader outlook is the New and True Chronology. It is a rigid and a correct scale of years, planted by the orbs of heaven, and in its light now certainly let in upon the Scriptures, Students of the Prophecies cannot err and the swiftest runner may discern the signs. It is to teach others the secret of this clairvoyance that we are now writing so that they, with us, may also see the waymarks as they speed along—*for not to see them henceforth is simply progress towards* 𝔇𝔢𝔰𝔱𝔯𝔲𝔠𝔱𝔦𝔬𝔫, and to see them as we do ourselves will force others irresistibly to swell *this* "MIDNIGHT CRY!"

Disavowing therefore that we "predict" anything, but fearlessly alleging that we believe "the end of the age" has come upon us, even as it has been circumstantially predicted by those who had the authority so to do, and being only desirous of imparting our own information, and the reasons for the faith that is in us, we admit the facts stated in this friendly letter and from the paper which accompanied it (The *National Tribune*, Washington, D C., Aug. 7, 1890,) extract the following]:

"It is estimated that one million Hebrews will have to leave Russia within the next few months, on account of the enforcement of the edict of 1882. According to this edict the Russian Jews must hereafter only reside in certain towns. *None of them will be permitted to own land, or to hire it for agricultural purposes*, and the order includes hundreds of villages which have large Hebrew populations. The Russian Jews *cannot hereafter have shares in or work mines. They are debarred from holding posts under the Government. They cannot enter the army, and will not be allowed to practice medicine, law, or to enter any of the professions.* Their residence must hereafter be confined to 16 of the provinces of Russia, and these provisions *will create an* EXODUS *of Israelites greater in number than the tribes who went forth under Moses.*" Verily before such bondage as is implied by the enforcement of an edict so abominable, that of Egypt itself pales to insignificance, nor is it a matter of wonder that the threat has begotten a concerted protest from the whole civilized world.

But Russia's policy is like that of the "Medes and Persians"—unalterable—and certainly, with fatuitous persistency, it lures her on to Armageddon. In the mean time, with Pharaonic lack of foresight, although paralleling Egypt's methods of oppression, and foreseeing its natural solution, the Czar does not intend to let oppression's children seek relief by exodus. Orders have been sent to the frontier customs posts, and to the railways in Poland, to watch for the threatened emigration of Hebrews from Russia!—(*Tribune.*)

In its editorial the New York *Tribune* says: "The Czar's infamous decree of religious intolerance excites much indignation in England, as it should in every country of the civilized world. Not since the dark ages has such a brutal spectacle been presented. It is not only a persecution of the Jews, but of all persons who do not conform to the Czar's own notions in religious affairs. All American and other Protestant ministers in Russia have been ordered summarily to quit work under pain of expulsion or worse. Even the Greek church itself is to be purged of all who are suspected of any degree of heterodoxy."

Truly we are upon the threshold of "the days of upheaval," and that *religious* elements are to be prominent factors in the solution of coming questions is no longer a matter of mere prophecy and prediction—they are already FACTS, and all "the East" is the caldron in which their seething issues are to boil. In Jeremiah's day the matter was a prophecy (Jer. i. 13-16), but in ours the ebullition has *begun*.

That we are not alone in these conclusions, but find them shared, in all their practical bearings, by those whom no one will accuse of being fanatics, pessimists, or alarmists, let us quote at length a late editorial of the New York *Tribune* (Aug. 10, 1890), for while we sit here working at our last appendix an ominous move has been announced upon the Eastern checkerboard. Under the heading of the "Sheik Against the Patriarch," the editorial is as follows:

"A few weeks ago two members of the Turkish ministry, Christians, resigned their portfolios, in protest against the Porte's unjust dealings with the Armenians. Now, Dionysius V, the Greek Patriarch at Constantinople, resigns his office in protest against the Porte's unjust dealings with the Greek Church. Standing alone, this latter event would be important. Regarded in connection with preceding events and with surrounding conditions, it is most ominous. This prelate is the head, it is true, of only one of the three great divisions of the Holy Oriental Orthodox Catholic Apostolic Church, and that by no means the largest. He has technically no authority over the Holy Synod of St. Petersburg, nor over the National Greek Church. Yet Constantinople, the city of the first Christian Emperor, is regarded as the center and capital of the whole Eastern Church, and so the Patriarch there has a certain sentimental supremacy over the heads of the other branches, and is in a measure the representative and spokesman of them all. He is, of course, the absolute spiritual head, and largely, too, the temporal head, of all the Greek Christians in European Turkey, who number not far from half of the entire population. It may easily be imagined, therefore, what a widespread sensation his resignation will cause, and how ill the Porte can afford to ignore such a protest against its policy.

"This resignation is one more unmistakable indication of the fanatical rule of the Sheik-ul-Islam over the Sultan and his ministers. This power behind the throne, backed by the Moslem priesthood and the mob, holds that the woes of Islam come because of compromise with sin, and that the only hope of the Empire lies in sternly smiting the infidel on every hand. So the Porte was driven to fasten new fetters upon the Christians of Crete.

"So it was constrained to turn a deaf ear to the cries of outraged Armenia. It was permitted to grant berats to Bulgarian bishops in Macedonia only because there seemed a chance of fomenting dissension between two branches of the Christian Church. Indeed, this very granting of the berats is the immediate provocation of the

Patriarch's resignation. But in other matters the Sheik-ul-Islam has driven the Porte to repeated acts of unjust discrimination against the Greek Church. And now, both in despair and in defiance, the chief prelate of that Church in Turkey resigns his office. The Patriarch has contended with the Sheik, and is overcome. So he practically makes an appeal to the country. For now every Greek Christian will take up the fight; and every Christian of any name, too, for by its conduct toward Greeks and Armenians alike, the Porte has shown its hostility to be directed toward no particular creed, but toward all who do not bow to Islam.

"It would be difficult to arouse a religious war in Western Europe. It would be easy to do so in Eastern Europe, where civilization is at a low ebb, and fanaticism is rampant, and where for centuries all sorts of oppression for conscience's sake have been perpetrated. The Greek and other Christians of European Turkey have suffered much because of their religion. Hitherto they have been almost helpless. But now they form a majority of the population, and they have the sympathy of other nations at their back. They see, moreover, the Ottoman power weaker than ever before. The Patriarch of Constantinople resigns, and thus rouses every Christian in the peninsula. Prince Ferdinand prepares to proclaim himself King of Bulgaria, and be no more a tributary vassal. Russia increases her army on the frontier, and demands $150,000,000 from an empty treasury. And the Sheik-ul-Islam, with the grip of blind fanaticism, holds the helm and keeps the Ottoman ship of state headed straight for the midst of the breakers. Truly, it is Kismet!"

Not a week passes, hardly a day, but that the "watchers" gather news like this and formulate it into ominous fulfillments of the ancient oracles —the only wonder is that even casual readers have not already had their eyes *forced open* to the import of the days ahead! But greater wonder is it, more than all, that those who are the "Shepherds" in Israel have not wakened to THEIR duty, and flung away the hypocrisy of their "higher criticism," and come out boldly for the truth "as it is written," and begun to *teach* the hungry flock that strays scattered on the hillsides.

May the Almighty judge them for the moments they delay, and bless the efforts of all who strive to herd the sheep into the pastures that are green.

In view of the discussion in this present volume, and of the conclusion arrived at, and already tabulated, that the true A. M. year commencing with this autumnal equinox marks the reversed parallel of the first year of Evil Merodach (II Kings, xxv. 27–30), the present news is very significant. But still more so is this fact stated in the *Tribune*, to wit: "Letters from Rabbis in Russia mention *September* as the period for enforcing the new anti-Jewish regulations." That is, the movement predicted in the Prophets, and now made clear in the present volume, will date officially with the beginning of a year, 5651 Jewish era, which we have already shown to be so *Tetragrammatal!*

The Washington *Tribune* states that "a convention of delegates, representing the largest colonies (of Jews) in 36 different cities of Russia, met, July, 10 months ago, and, after a long discussion, they gave *a unanimous vote for* PALESTINE *as their future home.*"

'The change that such an emigration would produce on the Holy Land cannot be overestimated. One million new workers would mean the adding of 150 per cent. to the population of the land occupied by ancient Palestine, and it would make Jerusalem a city of more than 100,000 people. The Holy City has been growing with almost American rapidity within the past few years, *and an exodus of the Jews from all parts of the world to it is slowly but surely going on.*"

What an impetus to such a movement this Russian Edict may engender, and what unknown increments may even now be shaking off the latency of suffering years no man can say, but all men who have read the Prophets with attentive minds, must rest convinced that they will be forthcoming in their proper season.

A close comparison of the two principal diagrams herein published ("The End of the Age" page 207, and general plan of the "Times of the Gentiles" page 213), will suggest the following dates as those which are to be most closely watched by all concerned: Sept., 1890; March and September, 1891; and March, 1892. The latter month will probably contain the most momentous event of history, and up to it, and certainly from it onwards until March, 1899, it will tax the modern Press, in telegraphic brevity, to keep apace of all that shall occur.

That outside of any such calculations, which all who are "wise" now agree are necessary in the premises, and of weighty import, that outside of them, and viewed from purely human standpoints, there is quite enough to warrant the concern with which the publications of the Our Race Company have endeavored to be heard, should be patent even to those who reject our premises. For instance, in an editorial upon the "Cost of 'Armed Peace,'" the New York *Tribune* of Sunday, Aug 3d, 1890, furnishes a most startling reminder of what from *very natural causes* will probably tend to bring about the state of affairs which we maintain, and have shown in this volume, was *supernaturally* predicted centuries ago. In view therefore of the peculiar prominence which our own studies cast upon the year 1892, the editorial referred to is very significant reading:—

"Although," it says, "the triple Alliance has undoubtedly contributed to preserve Europe during the last few years from the horrors of a general war, yet it is extremely unlikely that it will be prolonged beyond *January* 1892, *when the existing treaty expires* (!) Its continuation after that date is improbable, and even impossible, in consequence of the inability of either Italy or Austria to meet the financial burden which it imposes upon it.

And, again, "The stipulations have all been fulfilled to the letter, both by the Austrian and Italian governments, and will continue to be so until the present (7 year) compact lapses, at the end of next year. But the strain to which it has subjected Austria, and in particular Italian finance has been excessive—far more so, in fact, than is generally known. Neither of the two nations is in a position to submit to it any longer, and the Austro-Hungarian Minister of War, General Von Bauer, did not exaggerate the other day when he informed the Parliamentary delegations in tones of despair that the present state of affairs could not last. And, indeed, doubts have arisen in the minds of most the leading statesmen of the two countries in question, as to whether an attitude of conciliation and abnegation, rendered necessary by disarmament, would not be preferable to the disgrace of national bankruptcy, and to the ruin of national trade and industry."

But, there is a darker side to this inevitable rupture of the Triple Alliance—really the one at which the Alliance aims!—*the Russian side!* If, by dint of longer "staying powers" and by means of impenetrable "bluff," the great Northern Bear, aggressive by unbroken precedent, shall survive in arms an otherwise general European laying down of weapons, will she not have gained her point? Will the failure of the Triple Alliance to maintain itself, for pure want of means, alter the policy laid down in the "will of Peter the Great?" Is it to be reasonably supposed that the acute statesmen of Russia are ignorant of the extremities in which Europe finds herself, or will fail to profit by just that opportunity for which with consummate foresight they have been waiting? Europe is indeed in a serious dilemma—to disarm does not mean peace, to maintain the strain means Anarchy! In the meantime "the drift of affairs" in Turkey, says an occasional Vienna correspondent of the *Tribune*, is steadily, irresistibly and unmistakably towards the bad, and so bad has the condition now become that the end cannot be far off. The dismissal of Prince Bismark from office removed the only effectual guarantee of a peaceful settlement of this hopelessly involved Eastern Question, and the Porte, between the armies of Europe and the priest-led mob of Islam, between the devil and the deep sea, lets things drift to destruction as they please."

But here, too, Russia has a lien too long delayed in its foreclosure. The finances of Turkey are dried up, Russia knows it, and is now forcing the "Sick man's" hand. She has formally declared that the long delayed war indemnity must be paid at once, if not she will take forcible steps to collect it—

In reality she prefers the forcible means!

In the pending moments the ten great powers which are to be involved in the coming conflict are looming into view. In the west we have Portugal, Spain, France, Italy and Austria, the toes of one foot of Nebuchadnezzar's image;—in the east Egypt, Syria, Turkey and Greece, already form four toes of the other foot, and it seems almost certain that before the year is out the world will hear of King Ferdinand, of Independent Bulgaria, the missing toe! The image is unstable, the feet are formed of "iron mixed with clay"—of Monarchy and Anarchy!

The English, the Germans and the Russians never formed any part of the Roman Empire. Scotland and Ireland never saw a victorious Roman eagle, the true Britons were in Wales while Rome was present in the Islands of the West, and when Rome left Albion's shore, she not only, by two separate imperial edicts, officially absolved her from even nominal allegiance, but it was not until then that the never dominated Anglo-Saxon came in to stay!

It is ridiculous to count England into the "Image" of Gentile Sway!

Yet England will have much to say, and more to do in the coming stirring times,—for she it is who forms that other Empire—even the one of STONE! Germany and England, the Assyria and Israel of "the latter days," (Isa. xix. 23-25) are already significantly bound by a late secret compact,* and it requires but little perspicuity to discern the nation that will stand as *third* in such a Triple Alliance when the final day of need arrives!

But in this picture where does Russia stand? The answer is found in Ezekiel xxxviii. and xxxix. and where she stands she ultimately falls, and where she falls she lies!

However, ere this Northern Monster meets its final fate upon the slopes of Esdrælon she has a part to play, and in it is to be an incarnate agent of Evil.

But in the meantime there is a golden "hook" (Ezek. xxxviii. 4) which may yet check the anti-Jewish hostility of Russia for at least a spell,— *and Fate has strangely placed its control in Jewish hands!* We refer to the protest of the rich Jewish bankers of Paris, who have intimated that, unless the Czar at once guarantees the original *status quo*, they will join the German and English combination against Russian stocks!

---

* An offensive and defensive naval one, and it is believed even broader!

The Rothschilds have taken the initiative in this protest, and have followed it up so energetically that, through the good offices of the French government already in close alliance with Russia, it is believed they have gained a temporary point.

That such a forced restraint will change the aminus of Russia's treatment of the Jews is not to be considered for a moment—the spots upon the leopard will change first: nor probably will the impetus thus given to the Jewish Exodus be checked. We may therefore confidently look for an escape from bondage, and in due time for a parallel to the disaster which befell Pharaoh's hosts in the Red Sea.

And here, from sheer inability to follow this continually unrolling drama further, we must leave prognostications to events themselves. With others who have faith, we are content to await developments; and with them are satisfied that, no matter what unlooked for course they may eventually take, the lines laid down by Israel's prophets will be literally followed—nor delay beyond the times and seasons set for them of old.

Finally, dispatches from Berlin indicate that there was political significance in the late visit of the Kaiser, the outcome of which may astonish the world. It is believed that he is fully awake to the gravity of the European situation, and that the main object of his recent royal pilgrimages has been to bring about some new guarantee of peace. To this the Czar's assent is necessary, and the belief in official circles is that the Czar will treat the league of peace as a futility unless associated with the restoration of Russian dominance in the Balkans, and the permanent satisfying of France.

And thus it is that in the midst of warlike tension the cry of "Peace, Peace!" is raised, while the Bible has declared there "is no Peace!" That cannot be until the whole present system, with its broken faith and faithless guarantees is swept away. Next year is set for the universal peace convention. It is to meet in Rome, beneath the cracking dome of St. Peter's.* But whether it be fated to convene or not its deliberations can but be in vain, for so it is predicted.

The time has therefore come when men must see the things *as they are*, and as they have been written from of old, but not perhaps, before the outcome, nor from volumes such as this; yet none the less—must see them and believe.

Half a century ago there were only 3,000 Jews in the Holy Land, and there were but 32 Jewish families in Jerusalem. Now, about 40,000 out of

---

*The dome of St. Peter's has been cracking for a considerable length of time, and the number and extent of the fissures are becoming alarming. About a hundred years ago a similar state of things was remedied by encircling the dome with a strong band of metal. The band was heated, and its contraction on cooling was found to be sufficient to close up the cracks. The suggestion now made is that electric welding has come just in time to make St. Peter's safe for another hundred years.— The *Queries Magazine*, Buffalo, N. Y., August, 1890.

50,000 in the city of David are Jews, and if the Russian emigrants are allowed to enter,—and who, pray, shall contravene the prophecies of God?—the city will be built albeit it shall be "in troublous times."

Frank G. Carpenter, in his late article in the *National Tribune*, says: "When I visited Jerusalem about a year ago, I was told that three-fourths of the people were Jews, and I found Jews about the walls of the Temple of Solomon from every part of the world. The most curious among them were the Gaddites, a tribe which has lately come from the southern part of Arabia, and which has been shut off from the rest of the Jews of the world until now. These Jews had a number of rare manuscripts of the Holy Scriptures. They knew nothing of the New Testament, and had left Jerusalem (Palestine?) before Christ was born. They claim to have received a prophecy which warned them to come back to the land of their fathers, and they are now there tilling the soil. There are many Jews in Jerusalem from Morocco, and these are of such a character and belief that they have a strict class of their own. They are not under the protection of any European power, and they claim to have been in Spain at the time of the crucifixion. They state that they were driven from Spain by Ferdinand and Isabella during the same year that Columbus discovered America, and were forced to go with the Moors to Morocco. They speak Spanish, dress in oriental costume, and number at least 5,000."

But it is to no further purpose that we review the Jewish signs of the times, and those which concentrate upon the Eastern question. The threshold of the appointed days has certainly been reached, and from now on not single volumes, nay, not even libraries, may serve to compass all that could be written; but in closing, let one thing that has been written be repeated with concern to "Judah." In this series we are earnestly advocating the truth of Anglo-Saxon identity with Lost Israel—it is essential that Judah "walk with Israel" in the great return to the land of her ancestors. If she walks alone she is doomed to stumble, and if she mistakes her "Israel Ridivivus" she will stumble into doom!

There is one terrible prediction yet hanging over Judah's head—which, whether they believe in Him who uttered it or not, has all the force of nearly 1,900 years to lend it credence ere the day of its fulfillment.

"I am come in my Father's name and ye received me not, *if* another *shall come in his own name,* him *ye will receive!*" (John, v. 43).

The Saviour here refers to Anti-christ as the "Anarchos" whose advent is at hand!

C. A. L. T.

August 21st, 1890.

"*In the first year of Darius the son of Ahasuerus, of the seed of the Medes which was made king over the realm of the Chaldeans;*

In the first year of his reign, I, Daniel, understood by books the number of the years whereof the word of the Lord came to Jeremiah the prophet that he would accomplish seventy years in the desolations of Jerusalem." Dan. ix. 1-2.

PART III.

CHRONOLOGICAL TABLES.

"We account the Scriptures of God to be the most sublime philosophy.

"I find more sure marks of authenticity in the Bible than in any profane history whatever."
—Sir ISAAC NEWTON.

THE
"CHALDÆAN BABYLONIAN ERA,"

SYNCHRONIZED WITH BIBLICAL CHRONOLOGY

AND WITH

SECULAR HISTORY AND ASTRONOMY.

The object of the following tabular digest is to fix the ORIGIN of the Babylonian Era, in order that we may know its latter and terminal years with the most absolute accuracy.

The author submits it to the Christian world as a scientific demonstration, in so far as its astronomical and chronological references are concerned. It is but an extract from very voluminous notes. But as time will not permit, nor are means yet at hand to present its collateral chapters, it must stand in the present volume in a somewhat disconnected position.

If the Bible student will fairly examine it, against any and every reference contained in the Holy Writ, covered by the years that its own scope comprehends, he will find that it answers every special requirement, and so harmonizes the cross references as to demonstrate that its own inter-relations are without flaw.

As a preliminary step in the construction of this table it was demanded that every Biblical reference to the contemporary years of Kings and Chronicles and the Prophets and to the Hebrew Calendar, should be arranged agreeably to the record itself, without any modification whatsoever. The first result was, that so soon as the table had been thus completed, it was manifest at a glance that it was, *in esse*, a harmony—and one which needed no apology, anywhere along its sequence.

It was thereafter an easy matter to synchronize the scale thus resulting, with secular history, since, of course, many dates therein already synchronized, and the rest fell into line so soon as a few were fixed.

The astronomical and chronological data then clinched the scale against the A. M. years beyond any possibility of moving them. They will be found to agree with the rectified chronology already *viséd* by the British Chronological Society and now, for so many years published in "All Past Time."

In this table the B. C. years are the *true* ones, *i. e.*, they are reckoned from 3996 A. M.; to change them to Usher's B. C. dates, add 8 years, since his scale reckons from 4004 A. M.

PRELIMINARY

CHRONOLOGICAL OUTLINE.

3233 A. M. 763 B. C.

Iva-lush (Arbaces?) or the "Pul" of the Scriptures, an Assyrian General, governing the northern provinces, and Belesis, the governor of Babylon, having revolted, overthrew Nineveh and blotted out the "First Assyrian Empire."

With "Pul" the "Second Assyrian Empire" began and with Belesis the "Second Babylonian Empire." To anticipate now the thread of history this II Assyrian Empire, under Esar Haddon eventually put an end to the II Babylonian, and continued until Nabopolassar conquered it, and founded upon its ruins the *Chaldee* Babylonian empire.

The particular aim of this table is to settle the true opening year of this *latter* empire, to follow its first 66 years chronologically, and to fix this chronology beyond all peradventure by references to history, astronomy and the Bible.

3244 A. M. 752 B. C.

Menahem pays tribute to "Pul" the king of Assyria. 2 Kings xv. 19-20; 1 Chron. v. 3, 25.

3257 A. M. 739 B. C.

The edict of Nabonassar fixes his ERA, (Thoth 1 or Feb. 26th noon). This date is accurately determined by astronomical observations, including eclipses recorded by Ptolemy, the times of whose occurrencies were invariably measured by the ancient astronomers from it. From this year on to the 20th year of Artaxerxes (3557 A. M.), is just 300 years.

3263 A. M. 733 B. C.

"Within three score and five years shall Ephraim be broken that it be not a people."—Isa.vii. 8; *vide* 2 Kings. xvi. 5-8; " Lo-Ammi!"—Hosea i. 10. Isaiah's prophecy was uttered upon the Sabbath, the 6th day of the 6th civil month, (our February) and began to go into effect at once, since in this year Tiglath Pileser carried away "the Reubenites, the Gadites and the half-tribe of Manasseh." (1 Chron. v. 26).

3265 A. M. 731 B. C.

Tiglath Pileser II destroys Damascus. (western Asiatic Inscriptions of Brit. Museum, vol. iii p. 10) and Ahab visits him (2 Kings xvi. 9-16).

3282 A. M. 714 B. C.

The siege of Samaria, that is of the *City*, the capital of the land of Samaria, was opened by Shalmaneser, at the beginning of this year. It lasted 3 full years 3282-3-4. Shalmaneser died during the siege. He was succeeded by Sargon, who conducted it to its close. Its termination was in the 9th year of Hosea, and the 7th of Hezekiah. *Two* Samarias are referred to in 2 Kings, xvii. *i. e.* the city, verse 5, and the land, verse 6. The people of the land began to be moved away at once. Thus, however we regard the chronology of Chronicles and Kings, Hosea's actual reign over them, *i. e.* "Israel" was but nine years in duration. But this chronology is not at all in fault, a fact which is of sufficient importance to clear up at once, and so at last set right a place in Biblical exegesis which has always been a vantage ground for Infidels.

HOSHEA'S REIGN HARMONIZED.

NO. A.M.	HOSHEA.		AHAZ.	HEZEKIAH.
(1) 3274	1		12	2d Kgs. xvii.1
(2) 3275	2		13	
(3) 3276	½ ⎫		14	
(4) 3277	In 2 ⎬ 3d		15	2d Kgs. xvii. 4
(5) 3278	Prison ⎪ Yr.		16	
(6) 3279	½ ⎭ 4		xiii 1	2d Kgs. xviii. 1 / 2 Chron. xxix.
(7) 3280	5		2	3-17
(8) 3281	6		3	2d Kgs. xvii.5
(9) 3182	7		4	2 Kgs. xviii. 9
(10) 3283	8		5	2d Kgs. xvii.6
(11) 3284	9		6	2 K. xviii. 10

An examination of the foregoing table will show that there is no discrepancy between 2d

Kings xvii. 1, which refers to 3274 A. M.—the whole of which was Hoshea's "first" year,— and 2d Kings xviii. 10, which refers to 3284 A. M. —the whole of which was only Hoshea's "ninth" year, although this apparently makes his reign eleven years instead of "nine" in all. *Because*, there was an *interregnum* in his reign of exactly two years during which he was temporarily imprisoned by Shalmaneser (*vide* xvii. 4). After his release, he reigned two years—revolted in the third, and the siege of his capital began with his seventh of actual reign. A school-boy finding the Biblical data relating to this reign given as the parts of a fair chronological problem in his arithmetic would have solved it without difficulty, and by referring to chapter xviii. 1, would have obtained the key to the whole situation. *Half* of the "third" year here referred to was the first half of 3276 A. M. (The interregnum commenced with the first day (Monday) of the first sacred month of that year, and extended through the last day (Monday) of the sixth sacred month of the year 3278 A. M.) The remaining half of Hoshea's "third" year, therefore, commences at this point, Tuesday, 1st day 7th civil month, and ends with that civil year. But, and note this well, the year (3278 A. M.) was an intercalary one, the 12th of the Hebrew cycle. Thus its last, or 13th month is "*silent*," and here again bursts forth the consummate accuracy of the Biblical Chronology, and its all sufficiency when

PRELIMINARY OUTLINE. 147

faithfully handled. With the end of the 12th month Ahaz died—"that King Ahaz," (2 Chron. xxviii. 22) whose sacrifices to other gods than Jehovah "were the ruin of him and of all Israel" (23).—Hence his successor, Hezekiah, came to the throne of Judah on this same year 3278 A. M., in its 13th month, which was thus the last month of Hoshea's "third" year, and so the entire record, which has been such a knot to chronologists, completely tallies! Now this intercalary month of 3278 A. M. was the one in which Hezekiah "re-opened the doors of the House of the Lord and repaired them" (2 Chr. xxix. 3.) And here is then a fitting place, and text, whereat to call upon *Israel Redivivus*—at the beginning of this final intercalary period of one and one-half years, which (from the date of this volume to March, 1892, A. D.) intervenes before the closing week of the Babylonian era,—and to call as impressively as mortal man may be permitted, to re-open the doors of the House of the Lord and to repair them against his issue from within the veil. "Ecce venit," (1 Cor. xvi. 22) "Maranatha"—The Lord is Coming!

And finally, and in the meantime, here also is the place and occasion to challenge the world to point out one single case in the entire Chronology of His Word which its own unaided record and cross references are not sufficient to make plain to one that runs!

In the "intercalary months" the Hebrews avoided doing any sacred work, so far as possible. But this first act of Hezekiah was imperative, and yet it stopped at the opening and repairing of the doors themselves, and with the preparing of the Priests and Levites for the far more important task of cleansing the temple itself, (2 Chron. xxix. 4-11). This grander undertaking was begun upon the calendric "New Year's Day," on Wednesday the 1st day of the 1st civil month of 3279 A. M. The porch was reached on Wednesday the 8th, and thus the House was sanctified in eight days, "and on the 16th day of the first month" which was Thursday, "they made an end" (2 Chron. xxix. 17). And early upon the next day, Friday, the King arose (20) and all the rulers gathered to do sacrifice (21-28). Thus "when they had made an end of offering" it was *the Sabbath eve!* and "the King and all that were present with him, bowed themselves and worshipped" (29). "So the service of the house of the Lord was set in order, and Hezekiah rejoiced, and all the people, that God had prepared the people, for *the thing was done* SUDDENLY!" (36).

The Chronology of the Bible is simply marvelous, and the significance of this type should sink deeply into the hearts of those whom God is even now preparing for the antitypical task of cleansing his earthly temple for the final time.

But to return to our chronological outline; for quite different things were taking place at Samaria while the scenes we have now briefly sketched were being enacted at Jerusalem:—

3284 A. M. 712 B. C.

Samaria, the Capital, falls with end of year, *i. e.*, in its "Intercalary days." This is the usual date *à quo* for Israel's Captivity. It is not, however the one in which it was "Consummated" (*vide* 3317 A. M.); nor yet its most important *point d'appui* (*vide* 3306 A. M.

3285 A. M. 711 B. C.

The final Israelitish Captives leave Samaria as the year begins. They are first placed in Assyria. 2 Kgs. xvii. 5-23; (latter part of 6 refers to later events), also 2 Kgs. xviii. 9-12, except last part of verse 11.

Unless this work is astronomical it is not worth a straw chronologically. Two total lunar eclipses, agreeing with our modern ones (Nos. 48 and 50), must have occurred at Babylon upon the Sacred year which spans 3284-5 A. M. Now these two eclipses are actually recorded by Ptolemy as having so occurred, thus we have additional evidence that our "line of time" is continuous down to the last eclipse of history (June 17th, 1890).

150   THE BABYLONIAN ERA.

3292 A. M.   704 B. C.

Hezekiah's 14th year. Sennacherib ascends and reigns 24 years. He comes up against Jerusalem at once (2 Kings, xviii. 13–37, xix. 1–35). His army is destroyed in the closing days of this year; Hezekiah's sickness culminates, and Sennacherib returns to Nineveh (36). *Vide*, also, Isa. xxxvi. xxxvii.

3293 A. M.   703 B. C.

In the beginning of this, Hezekiah's 15th year, the shadow on the "DIAL OF AHAZ" turns back 10° (40 mts.), at "high noon" of Wednesday, the 18th day of 1st civil month, the sun being at that instant about to go into autumnal equinox. The solar year was therefore delayed, the day lengthened 40 minutes, and the calendar thus made absolutely correct (*vide* Joshua's Long Day, 2555 A. M.) Hezekiah's life was lengthened by 1 Calendric Cycle of 15 solar years, *i. e.*, to end of 3307 A. M.

3306 A. M.   690 B. C.

Sennacherib (and Esar-Haddon, his son, who was now jointly associated with him) having completed the conquest of Media, transfer "Israel" into its cities. Before the reign of Sennacherib Media was *unknown* ("B. M. Inscriptions," Vol. i. p. 63). This is the true date *à quo* of Israel's Captivity, 1st sacred month (7th civil), just 100 years before that of Judah. Its significance has been so completely hidden that the date has escaped all former chronologists. From it 720 years forward (*i. e.*, "double" 360, Jer. xvi. 18) lands us at 4026 A. M., when the Saviour was baptized, and was thus made the first "fisher" sent after "Israel." This year is also just 360 years after 2946 A. M., when Saul's sin culminated; and Israel's period of backsliding began. Students of Prophecy will at once perceive the significance of this date, March, 3306 A. M. Thence forward Israel's "7 times" of punishment expired in 5826½ A. M. (our 1828¼ A. D.), or 1260 lunar years (= 1222½ solar) beyond the celebrated "Decree of Phocas," in favor of Boniface III, the which date (607 A. D. common reckoning) is a very focus of prophetic *origines*.

Its discussion, here, is too intricate for such as are not supplied with prophetic "oil" (Matt. xxv. 1-13). Those who are may draw their own conclusions!

3316 A. M. 680 B. C

Sennacherib slain (2 Kgs. xx. 37), at end of year: Media revolts, her "era" begins, and Israel herself takes this occasion to escape through the "Gates of the Caucasus." Thus from 3285 to 3317 was less than half of "70 years," and from her actual location in Media to her escape was but one-seventh of Judah's term; it was indeed "a little while," for already had she begun her repentance (Jer. iii. 11; the whole of Hosea).

3317 A. M. 679 B. C.

Esar-Haddon ascended as sole monarch of Assyria at the beginning of the year, conquered Babylon (thus ending its "second" empire), and reigned 13 years. His second important work was the replacement of Israel by Cutheans, or "Samaritans," which was completed in his 12th year, 3328 A. M. Thus was Isaiah's prophecy of 3263 A. M. (q. v.) fulfilled, and as the 65th year ran out Israel's captivity was "accomplished"— "No more even a people;" she also had no empty land awaiting her at home (2 Kgs. xvii. 24-41, xviii. last part of 11).

3328 A. M. 668 B. C.

One of the most remarkable of the "connecting years" upon the Prophetic Scale. For instance, from its February there extend 2520 lunar years (2445 solar) to the accession, Feb. 15, 1775, of Pius VI, the Pope whose temporal government was overthrown by the French Revolution; this occasion, itself being 1260 "calendar" years from Justinian's Decree (Mar. 533 A. D.) making the Bishop of Rome, John II, "Head of all the Holy Churches and of all the Holy Priests of God."

From this same February of 3328 A. M. there extend 2520 solar years to the formal deposition of Pope Pio IX, Feb. 8th, 1849, the which date was itself 1260 "calendar" years from the Decree of Phocas conceding to Boniface III "the Headship over all the Churches of Christendom," and which latter was memorialized by the Pillar of Phocas at Rome bearing inscriptions and date.

3329 A. M.   667 B. C.

The 66th year from Isaiah's prophecy. Esar-Haddon dies at its termination.

3330 A. M.   666 B. C.

The 666th before Christ. Assur Banipal ascended throne of Assyria, reigned 41 years.

3371 A. M.   625 B. C.

Bel-zakir-iskum, or Assur-ebil-ile, ascends and reigns six years. With him in 3377 A. M. the "Second Assyrian Empire" ended. "He perished in the flames of his palace, which he fired, when the Babylonians and their allies entered Nineveh." *Vide* ("All Past Time.) Speaking of this final overthrow of Assyria, Labberton says: "It was one of the most terrible catastrophes that ever happened. Not only an empire was destroyed that a few years before had ruled the whole of Western Asia, but a whole nation, which for centuries had been the curse of all other nations was utterly effaced. The four capitals, Assur, Ninua, Kalach, and Dur Sarrukin, were so thoroughly blotted out that they never were inhabited again. They disappeared from the face of the earth as the nation that had built them." (Hist. Atlas, p. 12.) But upon the other hand, it will in due time be the task of the "*new* Chronology" to show that the "remnant of Assyria" followed "Israel" *West*, and that in the Germans they are yet to complete their history (Isa. xix. 23-25) vide "Our Race," study No. 1, *The Romance of History*, p. 167).

Having now arrived legitimately at the general date *à quo* of the "*Chaldee* Babylonian Empire," it remains for us to fix it still more positively by means of the most minute system of cross references. We shall therefore follow the next 80 years, one at a time, and show how rigidly their historical events are tied together, and to it.

THE ORIGIN OF THE BABYLONIAN ERA,
FROM WHICH WE MUST DATE
"THE TIMES OF THE GENTILES."

EVENTS AND REFERENCES.	Jos.	A. M.	B. C.
Josiah crowned :—1 Josiah *was* eight years old when he began to reign, and he reigned in Jerusalem one and thirty years.—2 Chron. xxxiv. 1.	1	3365	631
	2	3366	630
	3	3367	629
	4	3368	628
	5	3369	627
	6	3370	626
3 For in the eighth year of his reign, while he was yet young, he began to seek after the God of David his father.—2 Chron. xxxiv. 3.	7	3371	625
	8	3372	624
	9	3373	623
And in the twelfth year he [Josiah] began to purge Judah and Jerusalem from the high places, and the groves, and the carved images, and the molten images.— 2 Chron. xxxiv. 3. Date of Zephaniah and Habakkuk.	10	3374	622
	11	3375	621
	12	3376	620
Jeremiah commissioned; Nabopolassar father of Nebuchadnezzar revolts from Assyria, and ascends the throne of Babylon, 1st day of Sacred year, *i. e.*, 7th Civil month of **3377** A. M., this is the origin of the BABYLONIAN ERA.—Jer. i. 1.	Nab. -1	13	3377	619

N. B.—[To fix this Era, *i. e.*, its chronological origin as explicitly as possible, let us note as follows :—It is the middle of 3377 A. M. (*i. e.*, the beginning of its 7th Civil or 1st Sacred month). It commences *with* the 1296th year from Abraham's Call out of Ur, and *with* the 865th year out of Egypt, *i. e.*, from the Exodus. "Israel" had been *in* captivity 71 years when

Nabopolassar ascended the throne, and there extend from that event 29 full years to the date of "Judah's" captivity; *i. e.*, between "Israel's" and "Judah's" dates of captivity are just 100 years, no more, no less upon the Solar scale. Finally, this event, Nabopolassar's accession, was at the middle of the 848th *Lunar* year from Joshua's Long Day, or 822½ *Solar* years from this same date, and it was at the March Equinox of 3377 A. M. and 3377½ years of Solar astronomic *duration* from the "1st day" recorded in Genesis. It was also the middle of Josiah's 13th year. Jeremiah was "called" at the beginning of this 13th year, and so was prepared to see the full initiation of the "Times of the Gentiles."]

(3380 A. M. is the 483rd Sabbatic Year.)

IMPORTANT. — In the year 3381 A. M., a very nest of prophetic times concentrate. It is particularly important as marking a dense period of sin, both in Israel and Judah, and the one preceding the finding of the Law.

Nab.	Jos.	A. M.	B. C.
1–2	14	3378	618
2–3	15	3379	617
3–4	16	3380	616
4–5	17	3381	615

Ezekiel refers to this particular year in the first verse of his prophecies in a manner so occult that it has never before been explicable, because no other than the present and only accurate system of chronology can possibly compass its ramifying references. Thus, Ezekiel's book opens in the 4th Sacred month, the 5th day of the month, which was a Sabbath, of the year 3410 A. M., which was the 5th year of Jehoiachin's captivity (2). Nevertheless, in verse 1 he calls it the 30th year, and thus refers back to the year now under consideration, to wit: 3381 inclusive, *i. e.*, dating from the end of 3380 A. M. Now, the first 30 years of the 40 referred to in Ezek. iv. 6 commence here and run out in 3420, when Nebuchadnezzar took away his last batch of Jewish prisoners (Jer. lii. 30). But in the same chapter, iv., Ezekiel refers to 390 days (3-5) for "Israel," which typified years (6). From the year in which Ezekiel obeyed this command, namely, 3410 A. M., there extend backward 390 years to Solomon's sin (1 Kings xi. 1-25) which led to Jeroboam's revolt

TIMES OF THE GENTILES.

(26-40), and which was the original occasion of "Israel's" special sin, as detailed in 1 Kings xii. Solomon's defection and Jeroboam's sin and flight took place in 3020 A. M., and from thence *to* the year under consideration 3381 A. M., there extend just 360 years, or "One Time," which, with the 30 to Ezekiel's mimic siege make up the 390 referred to. This type and its own actual scale on the calendar of 3410-11 A. M., and the date as an *à quo* and *ad quem* refers to so many others, and verifies them, that it is impossible to devote more space to the matter here.

	Nab.	Jos.	A. M.	B. C.
8 Now in the eighteenth year of his [Josiah's] reign, when he had purged the land, &c., was—The law found (2 Kings xxii., 2 Chron. xxxiv. 14-33); The oath at "Bethel" taken (2 Chron. xxxiv. 29-32, 2 Kings xxiii. 1-3, xi. 14); and The Great Passover held (2 Kings xxiii. 1-23, 2 Chron. xxxv. 1-19); "In the eighteenth year of the reign of Josiah was this passover kept."	5–6	18	3382	614
69th and last year of Psammetichus I of Egypt.	6–7	19	3383	613
Pharaoh Necho ascends; reigns 16 years.	7–8	20	3384	612
	8–9	21	3385	611
	9–10	22	3386	610
	10–11	23	3387	609
	11–12	24	3388	608
	12–13	25	3389	607
	13–14	26	3390	606
	14–15	27	3391	605
	15–16	28	3392	604
Josiah slain by Pharaoh Necho middle of Civil and end of Sacred year.— 2 Chron. xxv. 20-27, 2 Kings xxiii. 29-30.	16–17	29	3393	603
	17–18	30	3394	602
	18–	31	3395	601

158   THE VOICE OF HISTORY.

EVENTS AND REFERENCES.	Nab.	Jeho-ahaz.	A. M.	B. C.
Jehoahaz succeeds at commencement of Sacred year and reigns but three months.—2 Chron. xxxvi. 1-3, 2 Kings xxiii. 31-34. 1 Then the people of the land took Jehoahaz the son of Josiah, and made him king in his father's stead in Jerusalem. 2 Jehoahaz *was* twenty and three years old when he began to reign, and he reigned three months in Jerusalem. 3 And the king of Egypt put him down at Jerusalem, and condemned the land in an hundred talents of silver and a talent of gold. 4 And the king of Egypt made Eliakim his brother king over Judah and Jerusalem, and turned his name to Jehoiakim. And Necho took Jehoahaz his brother, and carried him to Egypt.		−19	3395	601
Jehoiakim commences to reign the 1st of the 10th Civil, or 4th Sacred month.—2 Chron. xxxvi. 4, 5; 2 Kings xxiii. 34-37. 1 In the beginning of the reign of Jehoiakim the son of Josiah king of Judah came this word unto Jeremiah from the Lord, saying, 2 Thus saith the Lord to me, Make thee bonds and yokes, and put them upon thy neck.—Jer. xxvii. Jeremiah puts on his yokes this year, 3395, which was the IXth year of the Hebrew Calendar, and wears	−19	−1	ix. 3395	601

TIMES OF THE GENTILES.

Events and References.	Nab.	Neb.	Jeho.	A. M.	B. C.
them to the IXth of the next ; *i. e.*, 14 full years, and just into the 15th year. Jehoiakim pays tribute to Necho for three years commencing with— 1	–19		–1 *contd nued.* x.-xi.	ix. 3395	601
2	19–20 20–21		1–2 3396 2–3 3397	xi.-xii.	600 599
Nabopolassar dies in the middle of this year, and is succeeded by his son Nebuchadnezzar, whose first 3 three months of reign overlap the last three of Jehoiakim's 3d year, hence Daniel i. 1 is correct, as shown here.—Jer. xxxvi. 1-8 (3398-9).	21– –1	1	3– –4 3–	xii. 3398 xii. 3398	598 598
For a similar reason the following is correct : 1 The word that came to Jeremiah concerning all the people of Judah in the fourth year of Jehoiakim the son of Josiah king of Judah, that *was* the first year of Nebuchadrezzar, king of Babylon; 2 The which Jeremiah the prophet spake unto all the people of Judah, and to all the inhabitants of Jerusalem, saying,	1–2		4–5	xiii.-xiv. 3399	597

3 From the thirteenth year of Josiah the son of Amon king of Judah, even unto this day, that *is* the three and twentieth year, the word of the LORD hath come unto me, and I have spoken unto you, rising early and speaking, but ye have not hearkened.—Jer. xxv. 1-3.

(And so also is Jer. xxv. 3, since the 23 years begin at the beginning

160   THE VOICE OF HISTORY.

EVENTS AND REFERENCES.	Neb.	Jeho.	A. M.	B. C.
of 3377 A. M. and run to the end of 3399 A. M., *i. e.*, inclusive.) Jer. xlv.—Baruch comforted. Pharaoh-necho, smitten by Nebuchadnezzar, is succeeded by Psammetichus. "Pharaoh-necho king of Egypt, which was by the river Euphrates in Carchemish, which Nebuchadrezzar king of Babylon smote in the fourth year of Jehoiakim the son of Josiah king of Judah.—Jer. xlvi.	1–2	4–5	xiii.-xiv. 3399 *conti nued.*	597
Jer. xxxvi. 9–32—Jehudi burns the Roll (Dec.). Nebuchadnezzar's vision.—Dan. ii.	2–3	5–6	xiv.-xv. 3400	596
Third and final year of Jehoiakim's tribute to Nebuchadnezzar. Last year (xv.) of the Calendric Cycle. Historical eclipse, Central Solar, No. 1 of the Team [date on Usher system B. C. 603 (*i. e.*, before 4004 A. M.)] True date, 3401 A. M.; last repeated, Jan. 22, 1879. N. B.—This same sequence of eclipses passes through that of June 17th, 1890, and March 29, 1112. It also verifies the Beth Horon conjunction, and the equinox at High Noon marked by a reversed shadow upon the Dial of Ahaz alluded to in the text.	3–4	6–7	xv. 3401	595
Jehoakim rebels.—2 Kings xxiv. 1. Hebrew Calendric Cycle commences.	4–5	7–8	i.-ii. 3402	594

TIMES OF THE GENTILES. 161

Events and References.	Neb.	Jeho.	A. M.	B. C.
	5-6	8-9	ii.-iii. 3403	593
	6-7	9-10	iii. 3404	592
Pharaoh Hophra ascends, reigns 25 years.	7-8	10-11	iv. v. 3405	591
Jehoiakim captured by the Babylonians (2 Chron. xxxvi. 6), and dies a prisoner (2d Kings xxiv. 6), having reigned to the 4th quarter of the Sacred year. Transit of Mercury and recommencement of his 38th team of 15 transits each. The transit of 3406 was the 570th transit. It was 405 transits ago, reckoning from its future repetition in 1891 A. D.	8-	11-	v.-vi. 3406	590
His son Jehoiachim succeeds and reigns three months (the last quarter of the Sacred year) and ten days into the 1st Sacred month of the next Sacred year (2 Chron. xxxvi. 9-10). Hence "at the return of the year" Nebuchadnezzar sent for him (2 Chron. xxxvi. 10). The city was smitten on Thursday, the 9th day of the 1st Sacred month. Jehoiachin went out to the Babylonians on Friday, the 10th (2 Kings xxiv. 12), and the Captivity of Judah began. It being still the 8th year of Nebuchadnezzar just ending (12).	8-	Jehoiachin.	v.-vi. 3406	590

Events and References.	Cap.	Neb.	Zed.	A. M.	B. C.
Nebuchadnezzar now places Zedekiah upon the throne (2 Chron. xxxvi. 11, 2 Kings xxiv. 17-20, Jer. lii. 1-4.) Zedekiah's years thus run with the Sacred calendar, and lag a little, *i. e.*, they commence not earlier than Sabbath, the 11th day of 1st Sacred month of 3406, *i. e.*, with the Captivity, nor later than the Passover of that year, a very probable date in view of the solemn oath he is known to have taken to Nebuchadnezzar before the Lord upon his accession.	-1	-9	-1	v.-vi. 3406	590
And it came to pass the same year, in the beginning of the reign of Zedekiah king of Judah, in the fourth year, *and* in the fifth month, *that* Hananiah the son of Azur the prophet, which *was* of Gibeon, spake unto me in the house of the Lord, in the presence of the priests and of all the people, saying.—Jer. xxviii. 1-14. The expression "the same year"	1-2 2-3 10-11 3-4 11-12	9-10 2-3 3-4	1-2 2-3 3-4	vi. 3407 vii.-viii. 3408 viii.-ix. 3409	589 588 587

means of the *Cycle*; compare reference above, opposite 3395 A. M. The IXth year was just again *beginning*, and it overlaps 22 days into 3409 A. M., because the intercalation was not due until the end of the next year, when the Lunar year was "floated" thereby on to Solar time. This was what the false prophet Hananiah meant by saying "within two full years" (3). He really meant to those who understood their own calendar, "within the time that two full years 'float' together," or

TIMES OF THE GENTILES. 163

"before the intercalary days," &c. Scripture is filled with this and similar expressions, and they always refer to the intercalary year. In the present instance this was 13 months forward. Now, owing to the fact that the months of Zedekiah's reign begin with the 11th day of the corresponding Sacred month it follows that the 5th month referred to in verse 1 (Jer. xxviii) extended into the VIth Sacred month 11 days, while, as we have just seen the IXth year of the calendar overlapped 22 days into this same month from the other direction. There were just 30 days in the VIth Sacred month, hence the three last days of the fifth month referred to in the text were in reality the 9th, 10th and 11th days of the regular VIth Sacred month, and were also the 1st three days of the IXth year of the *Sacred* calendar backed up (as it were) to meet them. They were Tuesday, Wednesday and Thursday, upon either one of which the events detailed in verse 1 may have occurred ; but, so close is the limit, upon no others. The day was probably the central one or Wednesday, the 10th of the VIth Sacred, which was the 28th day of *Jehoiakim's* "Vth" month. This is, perhaps, one of the most beautiful instances that can be cited in the general chronology of the Bible, as an example of consummate accuracy, combined with a suggestiveness which is its own commentary upon the context. Had it not been carefully recorded that Jehoiakim reigned three months *and 10 days*, and had we not rigidly considered these days in the chronology, our work would here find itself in a hopeless *cui de sac* in the vain effort to make a IXth year of the cycle synchronize at all with a 5th month of Zedekiah's 4th year. If now we suppose that Zedekiah's ascension, or at least his oath to Nebuchadnezzar, dates from the Passover of 3406, we may run the splice down 4 days more, or make it cover a whole week. The 13th day of the month thus covered was the Sabbath preceding the Passover of 3409 A. M., upon which these incidents could even more fittingly have taken place. It is now possible to obtain a clear idea of the return judgment passed by the prophet Jeremiah upon Hananiah, when he had broken the yoke that the former had been wearing so long.

15 Then said the prophet Jeremiah unto Hananiah the prophet, Hear now, Hananiah ; The Lord hath not sent thee ; but thou makest this people to trust in a lie.

16 Therefore thus saith the LORD ; Behold, I will cast thee from off the face of the earth : this year thou shalt die, because thou hast taught rebellion against the LORD.

17 So Hananiah the prophet died the same year in the seventh month.—Jer. xxviii. 10-17.

The seventh month of the IXth year of the cycle was the XIIth month of Zedekiah's 4th year. Hence, in either case, it was "this same year."

	Cap.	Neb.	Zed.	A. M.	B. C.
Ezek. i. 1-2. 4th Sacred month, 5th day, Sabbath. (See chapter iv. Ezekiel.) The prophet receives his command on Sabbath the 5th, takes a week for his preparations, rests Sabbath the 12th and commences his "seige" of the "tile" on Sunday the 13th of the 4th month, 3410. Thence 430 days expired with Monday the 23rd of the 5th Sacred month, 3411 A. M., leaving him 7 days (a week) in that month to purify himself according to the law. Five days after, Sabbath 5th of 6th month, 3411, we, therefore, find him (Ezek. viii. 1) in his own house, and the recipient of his second vision. A consultation of the calendars given in the appendix to No. 3 will verify this and all other knotty points of the Scriptural chronology.	4-5	12-13	4-5	ix. 3410	586

TIMES OF THE GENTILES. 165

Events and References.	Cap.	Neb.	Zed.	A. M.	B. C.
(At the end of this year the intercalary month floated the Lunar and Solar years so as to re-commence together in 3411. This is very important as otherwise the 430 days above referred to would not come out. Ezek. i. 1 does not refer to 3411, but to 3410 A. M., as does also 2; Chaps. i.-vii. inclusive are parts of the same vision.)	4–5	12–13	4–5	ix. 3410 *continued.*	586
"The Sixth year, the 6th month, the 5th day of the month" (Ezek. viii. 1), also a Sabbath, and near the end of the Civil year 3411. Ezekiel's dates are always Captivity years and Sacred months (so, too, the *years* of the Exodus regarded as *units* of an "Era" date from the first of the Sacred month, although the Exodus itself was upon the 15th day of that month).	5–6	13–14	5–6	x.-xi. 3411	585
Ezek. xx. 1, "The Seventh year, in the fifth month, the tenth day of the month," Sabbath. (It must be noted that the Sacred years employed by Ezekiel began 10 to 15 days earlier than Zedekiah's years of personal reign.)	6–7	14–15	6–7	xi.-xii. 3412	584

THE VOICE OF HISTORY.

Events and References.	Cap.	Neb.	Zed.	A. M.	B. C.
	7–8	15–16	7–8	xii. 3413	583
	8–9	16–17	8–9	xiii.-xiv. 3414	582
10th Sacred month, 10th day (4th Civil month)—Ezek. xxiv. 1, 2, Sunday. Jer. lii. 4, Nebuchadnezzar lays siege to Jerusalem.		9–10	17–18	9–10	xiv.-xv. 3415	581
(Jer. lii. 29) Ezek. xxix. 1, 10th mo., 12th day, Sabbath.	10–	18–	10–	xv. 3416	580
Ezek. xxvi. 1, 1st month, 1st day, Sabbath. Ezek. xxx. 20, 1st month, 7th day, Friday. Ezek. xxxi. 1, 3rd month, 1st day, Tuesday. Jer. lii. 5-11, 4th month, 9th day "city broken up," Friday. Jer. lii. 12-24, 5th month, 10th day, Sunday. Temple burned.	–11	–19	–11	3416	580
				Siege of Tyre.	
Jer. xliii, beginning of the Civil year and cycle.	11–12	19–20	1	3417	579
Ezek. xxxiii. 21, 10th month, 5th day, Sabbath.	12–13	20–21	2	3418	578
Ezek. xxxii. 1, 17, 12th month, 1st day and 15th day, Fridays.					
Historical eclipse, central solar, No. 1 of team. Successive to one noted in 3401 q. v. 70 eclipses then as now between the two. No alteration of the length of day, hour, lunation, precession, seasons, or times.	13–14	21–22	3	3419	577

TIMES OF THE GENTILES. 167

Events and References.	Cap.	Neb.	Siege of Tyre.	A. M.	B. C.
Sequence of the days as rigid as the word of God. No lapse in the week. The chronology of the Bible is agreeable to chronological astronomy to the very last ultimate of time!)	13–14	21–22	3 continued.	3419	577
Jer. lii. 30, End of Ezekiel's 40 years for "Judah." Vide Ezek. iv., also see remarks opposite A. M. 3410 and 3381.	14–15	22–23	4	3420	576
	15–16	23–24	5	3421	575
	16–17	24–25	6	3422	574
	17–18	25–26	7	3423	573
	18–19	26–27	8	3424	572
	19–20	27–28	9	3425	571
	20–21	28–29	10	3426	570
	21–22	29–30	11	3427	569
	22–23	30–31	12	3428	568
Jer. xliv. In this year Tyre fell, and Hophra's Lybian expedition failed. The revulsion of feeling in Egypt leads to revolt under Ahmes.	23–24	31–32	13	3429	567
Nebuchadnezzar's first invasion of Egypt. Ahmes in meanwhile slays Hophra. Nebuchadnezzar confirms him as Amasis. In the confusion Jeremiah and the Royal Remnant disappear. Ezek. xl. 1, 1st month, 10th day, Thursday. (This vision of	24–25	32–33	−1 Ahmes.	3430	566

Events and References.	Cap.	Neb.	Ahmes.	A. M.	B. C.
Ezekiel covers the remainder of his Book of Prophecy and is the last date given by him. It extends to v. 35 chapter xlviii.) The "self same day" of Ezek. xl. 1, refers to the Thursday the 9th day of 1st Sacred month of 3406. When the City was "smitten," Jehoiachin went out on the 10th day, Friday, and as above noted Zedekiah may have begun to reign anywhere from the 11th (Sabbath) to the 14-15th (Passover) of that month. From the accuracy with which these cross references all come out upon the true chronological scale it must be manifest how *precise* the Bible is in all its records. The year 3430 is a most important one.	24–25	32–33	–1	3430 *continued.*	566
Nebuchadnezzar returns to Babylon.	25–26	33–34	1–2	3431	565
Ezek. xxix. 17, 1st month, 1st day (Tuesday).	26–27	34–35	2–3	3432	564
Daniel iii. The Image, and the Fiery Furnace. The significance of this act of Nebuchadnezzar is thus apparent. Thirty-four years had now elapsed since his "vision of Empire," and its significant interpretation by	27–28	35–36	3–4	3433	563

TIMES OF THE GENTILES. 169

Events and References.	Cap.	Neb.	Ahmes.	A. M.	B. C.
the Prophet Daniel ii., in 3400 A. M. In the meantime Palestine had been subjugated, Tyre reduced, and Egypt was now at last beneath his heel! The glory was too much even for that "Head of Gold." So, turned with pride, he was beside himself and reared an image unto vanity. The incident of the fiery furnace, however, brought him once more to short lived reason, and led to his re-acknowledgment of God's decree (Dan. ii. 28-30). It was now necessary to repeat the lesson in the form of a vision, so at the beginning of the next sacred year and of his own 37th year—	27–28	35–36	3–4 *contnued.*	3433	563
Nebuchadnezzar dreams of a tree. Dan. iv. 4-27. His 2nd invasion of Egypt now undertaken. It covers the 37th year of Nebuchadnezzar. His armies went from Migdol to Syene, and clothed themselves with Egypt's spoils. Our only monumental inscription of Nebuchadnezzar, referring to his wars, a clay tablet now in the British museum,	28–29	36–37	4–5	3434	562

Events and References.	Cap	Neb.	Abmes	A. M.	B. C.
absolutely confirms this 2nd expedition, and the year in which Nebuchadnezzar undertook it. *Vide* Trans. Soc. Bib. Arch., vol. vii. pp. 210-225; Vigoroux, vol. iv., p. 376. (N. B.—This 2nd invasion was three years after the first.)	28–29	36–37	4–5 *conti*|*nued.*	3434	562
Nebuchadnezzar returns to Babylon in time to be there at the beginning of the 7th Civil month, which was 12 months after his dream of the tree, and the incidents related in Dan. iv. 28-33 took place. His renewed victory in Egypt had again proved too much for even such an intellect, and this time it fell with a crash. He was insane from the middle of 3435 to the middle of 3442.	29–30	37–38	5–6	3435	561
	30–31	38–39	6–7	3436	560
	31–32	39–40	7–8	3437	559
	32–33	40–41	8–9	3438	558
	33–34	41–42	9–10	3439	557
	34–35	42–43	10–11	3440	556
	35–36	43–44	11–12	3441	555
Dan. iv. 34-37. Nebuchadnezzar's understanding returns. During the remainder of this year his kingdom is restored to him, and his first act is to write	36–37	44–45	12–13	3442	554

TIMES OF THE GENTILES.

Events and References.	Cap.	Neb.	Ahmes.	A. M.	B. C.
an epistle declaring his experience to all nations. *Vide* Dan. iv. 1-3, 4-27, 28-33, 34-37. And at the end of the year he died.	36-37	44-45	12-13	3442 conti- nued.	554
The 66th year of "the Times of the Gentiles!" Evil Merodach ascends the Babylonian throne at the beginning of this Civil year, the first half of which was thus the latter half of the 37th year of the Captivity. Hence Jer. lii. 31-34, and 2 Kings xxv. 27-30, are absolutely right. This entire year, 3443 A. M., marks a "lifting up" era in Judah's affairs, and has a marked bearing (typically) upon the year A. D., which fully counterparts it from *this* end of the scale (to wit, 5890 A. M., or from Sept. 1890, to Sept., 1891). The half year which follows Sept., 1891, corresponds by reversion to that in which Nebuchadnezzar's reason was restored, and he wrote his epistle to the nations. It is the firm conviction of the writer that from this coming September these counterparts will evolve in striking incidents!	37-38	1	13-14 Evil Mer'd'ch	3443	553

Events and References.	Cap.	Evil Mer'd'ch	Ahmes.	A. M.	B. C.
This year, 3444 A. M., marks the beginning of the 8th team of Transits of Venus (8 in a team). The transit occurred in Dec. 3444, and was repeated five teams later on Tuesday, Dec. 8-9th, 1874, which was also new moon's day. There are always 486 years between similar transits. From the beginning of this year to the end of 5888 A. M. (our Sept., 1890), there extend 2445 Solar years=to 2520 exact lunar, calendric, or "shortened" years. After which 1½ years of "silence," or rest, or warning, or Jewish irredentalism preceed and usher in the last and dreadful 7 final years of the Babylonian Era, to wit, those of ANTICHRIST.	38-39	2	14-15	3444	552

"HERE IS WISDOM.

Let him that hath understanding *count* the *number* of the beast; for it is the *number* of *a man*; and his number is **Six Hundred Threescore and Six.**"—Rev. xiii. 18.

To accept this mark (whatever it may be) is to sin against the Holy Ghost!!! 1 John v. 16. Matt. xii. 31-32. Rev. xiii. 16; xiv. 9-11. Mark iii. 28-29. Luke xii. 10. Heb. vi. 4; x. 26-29.

THE
CHRONOLOGY OF JEREMIAH,
FROM
HIS "CALL" TO HIS "DISAPPEARANCE,"

VINDICATED.

"*See, I have this day set thee over the Nations and over the Kingdoms, to root out, and to pull down, and to destroy, and to throw down, TO BUILD and TO PLANT.*" Jer. i, 10.

JEREMIAH VINDICATED.

It is at last high time to do justice to human faith, and to have it put on record, in at least one religious book, that we believe in the plenary inspiration of Jeremiah, and in that of the whole college of the Prophets, and that also we believe Jehovah is not only unswerving in his "times and seasons," but has so arranged them as to be within the understanding and demonstration of his creatures.

The whole library of Scriptural Commentary will be searched in vain to find a due recognition of the character and mission of Jeremiah, and we do not hesitate to arraign the entire Christian Church, heretofore and hitherto, as solely responsible for the deadly heresy of infidelity which feeds upon its own halting and apologetic volumes. No words are strong enough to paint the nature of a sin at once so fatal, and so suicidal to the integrity of faith, as has resulted from this willful disbelief in God's commands and promises to Jeremiah, and we challenge the orthodox of any branch of the so-called "church" to produce a single volume from the whole library of accepted standard theology wherein the doctors have recognized the *necessity* of as duly accounting for "*the* BUILDING *and* PLANTING," as for the merely historical bolstering up of a few minor incidents in this prophet's life.

Their whole teaching is at fault, in that, on coming to the end of Jeremiah's own works, so far as *human* history goes, and finding therein no mention of his having built or planted, they have tacitly and inconsistently suffered it not only to be implied, but have actually taught that he *failed!*

The inconsistency lies in the fact that if he *did* fail, then, *ipso facto,* his orders and authority were *not* from JEHOVAH, and hence his book would be without canonical value. And it is manifest that whether they teach this or not, the common sense of other men must pronounce this judgment upon it, or else reject their methods, and re-commence the study in a different spirit.

This is the only logical estimate to be placed upon the sacred works of Jeremiah; for if no more can be said of him than the college of his commentators have said, they have but helped "the world" to disprove the word of God!

But how grievous an error this has been, and is, if still persisted in, it has been the task of those who believe in the Israelitish origin of "Our Race" to show, and it shall be our own task to add another chapter to this prophet's vindication.

To commence: let it be pointed out, that it seems to have escaped the understanding of the Doctors, that Jeremiah was *Called,* and *Commis-*

sioned by the Holy Spirit at a moment which purposely ante-dated the *à quo* of Gentile times, just long enough to make him literally the "Prophet of the Nations," and to enable him to sweep into his philosophy the whole compass of their "times and seasons."

As we have already shown (Romance of History, p. 172) Daniel's Book closes at the date *à quo* the scattering commenced. Thence to Egbert's accession, we have already shown, is 1335 years upon the broader scale of prophecy. But Jeremiah's book *opens* at an equally significant era, to wit: the beginning of the year 3377 A. M., at the Passover or Spring-tide, of which Nabopolassar began the BABYLONIAN ERA. The momentous significance of this fact, now for the first time raised to the prominence it deserves in Chronology, must be apparent to all earnest students of the Bible.

In the meantime, before we begin to study the chronology of Jeremiah, let us point out to those who find so much difficulty in the fact that the Saviour, and others, referred to texts not now found in "Jeremy, the Prophet," (Matt. xxvii, 9, and 2 Chron. xxxv. 25), will perhaps recover their *judgment* by referring to 2 Maccabees, ii, 1, and Jer. xxxvi. 32, from which, together with Jer. xxxii. 12-14, it will be manifest that WE, moderns, have not the *whole* of his works in our possession, and that some of them are still buried against a day of great future necessity!

Let us now devote a few pages to the Chronological re-arrangement of the early part of the book of Jeremiah, in order that, with cleared insight, we may mentally follow the sequence of events into which they eventuated, and out of which they sprang.

A study of this description is like the solution of a problem in mathematics, and must be conducted to a rigid verification without neglecting any of the conditions involved. God's word does not admit of approximations as to fulfillment, each jot and tittle in the sequence of events must fall into its appropriate place, and until such a solution is arrived at, as shall thus harmonize the whole, and bind it into one complete history, it stands to faith, and to reason soundly weighing the premises, that the problem cannot have been integrated. It was in this spirit that the present solution was undertaken, and not until after many trials had failed to satisfy the earnest search for the truth, the whole truth, and nothing but the truth, did the final result come out in perfect chronological harmony.

PRELIMINARY CHRONOLOGICAL OUTLINE.

At the beginning of Josiah's 13th year, which was that of 3377 A. M., Jeremiah, then a mere youth, received his CALL as the "Prophet of the Nations." His "Commission" is broadly summed up in the first chapter of his own prophecies, and shows the scope of what he was, *personally*, appointed to review, and of which he was to be a part. In this first chapter he himself details his "Call" as follows:

THE BOOK OF THE PROPHET

JEREMIAH.

CHAPTER I.

1 The time, 3 and the calling of Jeremiah. 11 His prophetical visions of an almond rod and a seething pot. 15 His heavy message against Judah. 17 God encourageth him with his promise of assistance.

The words of Jeremiah the son of Hilkiah, of the priests that *were* in Anathoth in the land of Benjamin:

2 To whom the word of the LORD came in the days of Josiah the son of Amon king of Judah, in the thirteenth year of his reign.

3 It came also in the days of Jehoiakim the son of Josiah king of Judah, unto the end of the eleventh year of Zedekiah the son of Josiah king of Judah, unto the carrying away of Jerusalem captive in the fifth month.

4 Then the word of the LORD came unto me saying,

5 Before I formed thee in the belly I knew thee; and before thou camest forth out of the womb I sanctified thee, *and* I ordained thee a prophet unto the nations.

6 Then said I, Ah, Lord GOD! behold, I cannot speak: for I *am* a child.

7 But the LORD said unto me, Say not I *am* a child: for thou shalt go to all that I shall send thee, and whatsoever I command thee thou shalt speak.

8 Be not afraid of their faces: for I *am* with thee to deliver thee, saith the LORD.

9 Then the LORD put forth his hand, and touched my mouth. And the LORD said unto me, Behold, I have put my words in thy mouth.

10 See, I have this day set thee over the nations and over the kingdoms, to root out, and to pull down, and to destroy, and to throw down, TO BUILD, AND TO PLANT.

11 Moreover the word of the LORD came unto me, saying, Jeremiah, what seest thou? And I said, I see a rod of an almond tree.

12 Then said the LORD unto me, Thou hast well seen: for I will hasten my word to perform it.

13 And the word of the LORD came unto me the second time, saying, What seest thou? And I said, I see a seething pot; and the face thereof *is* toward the north.

14 Then the LORD said unto me, Out of the north an evil shall break forth upon all the inhabitants of the land.

15 For, lo, I will call all the families of the kingdoms of the north, saith the LORD; and they shall come, and they shall set every one his throne at the entering of the gates of Jerusalem, and against all the walls thereof round about, and against all the cities of Judah.

16 And I will utter my judgments against them touching all their wickedness, who have forsaken me, and have burned incense unto other gods, and worshipped the works of their own hands.

17 Thou therefore gird up thy loins, and arise, and speak unto them all that I command thee : be not dismayed at their faces, lest I confound thee before them.

18 For, behold, I have made thee this day a defenced city, and an iron pillar, and brasen walls against the whole land, against the kings of Judah, against the princes thereof, against the priests thereof, and against the people of the land.

19 And they shall fight against thee ; but they shall not prevail against thee ; for I *am* with thee, saith the Lord, to deliver thee.

This single chapter compasses the whole "Times of the Gentiles," and from the 11th to the 17th verses, its significance is still future—aye, even in OUR OWN *immediate future, i. e.*, must terminate with this current century!

But all the preceding verses, 1-10, concerned Jeremiah himself. In this connection, the last paragraph—"*to build, and to plant*"—of verse 10, deserves special attention.

Finally the closing verses, 17-19, amount to an explicit guarantee of personal immunity from all serious bodily harm to Jeremiah himself. In view of them it is manifest that all the legends with which his disappearance has been attributed to eventual martydom, current among Jews and Christians, are utterly baseless. It is a choice between Jehovah's promise, and human ignorance, and there should be no hesitation, so far as religious teachers are concerned, as to which side to support.

In the middle of the year of Jeremiah's "call" Nabopolassar, who was governing Babylon as an Assyrian province, revolted and was crowned as the independent king of Babylon. The date of his accession synchronizes with the beginning of the Sacred year (7th month of 3377 A. M.), and with it the "Babylonian Era" commences. Nabopolassar reigned thus independently for 21 years, and was succeeded by his son, Nebuchadnezzar, in the year 3398 A. M., whose own years *lag* a little on the sacred calendar.

Previous to the accession of Nebuchadnezzar, Jehoiahaz had been dethroned by Pharaoh Necho (3395 A. M.), and taken to Egypt, and his brother Jehoiakim placed upon the throne of Judah. The latter paid tribute to Necho until the first year (3398 A. M.) of Nebuchadnezzar, who then "smote" Necho, and *ended* his reign (Jer. xlvi. 2). For three years thereafter Jehoiakim was a faithful tributary to Babylon, but in his 8th year he rebelled, and renewed his Egyptian alliance, Psammetichus II being the reigning Egyptian Pharaoh, and then in the 6th or final year of his reign. These circumstances forced the Babylonians to renew their attention to Syrian affairs, and brought about the situation with which we shall commence this preliminary digest.

Incident upon the rebellion of Jehoiakim, and the prospective alliance of Judah and Egypt, came the accession of Pharaoh Hophra to the throne of the latter kingdom, in 3405 A. M., and the first half of this calendric year, which marked the closing months of Nebuchadnezzar's 7th year, found the armies of the latter marching into Palestine in order successfully to complete several tasks.

Early in the next year, 3406 A. M., Jehoiakim having reigned 11 years, was quickly overthrown by the bands of Babylonians and their allies (2 Kgs. xxiv. 2), and died a prisoner (2 Kgs. xxiv. 6) in his fetters (2 Chron. xxxvi. 6), before he could be sent to Babylon with the rest of the captives.

In the meantime (3405) Nebuchadnezzar had devoted his personal attention to the Egyptian complications of the situation, had cleared the land throughout its length and breadth of their armies, and having finally succeeded in forcing them back into their own country, had left the borders heavily patrolled (2 Kgs. xxiv. 7). It was soon after this juncture that Psammetichus II was succeeded in Egypt by Pharaoh Hophra.

The death of Jehoiakim quickly followed. This latter event naturally resulted in the succession of his own son, Jehoiachin, then 18 years old, to the throne of Judah, and drew the personal attention of Nebuchadnezzar, now free from all Egyptian complications, particularly to the city of Jerusalem. He therefore (3406 A. M.) came up against it immediately (2d Kgs. xxiv. 10, 11).

Jehoiachin's reign covered the last quarter of the current sacred year, and ran 10 days into the 1st month of the next; all of which was covered by the latter part of Nebuchadnezzar's eighth year, it being now the middle of 3406 A. M. The renewed siege of the city seems to have covered this entire reign of Jehoiachin, and in general terms the Babylonian operations in Syria, now under consideration, occupied the last half of the 7th and the whole of Nebuchadnezzar's 8th year of personal reign.

The chances of Jerusalem, which were hopeless from the start, now reached their crisis, and, as the sacred year drew to its close, Jehoiachin himself made up his mind to surrender. The new sacred year commenced, and with its return Nebuchadnezzar sent his ultimatum to him. The city was smitten or broken up on Thursday the 9th, and upon Friday the 10th day of this first sacred month, Jehoiachin went out and surrendered unconditionally. The next day, Sabbath, 11th, of 1st sacred month, 3406, thus marks the beginning of the Captivity, *i. e.*, its first day's ending.

The king of Babylon now made Mattaniah, the son of Josiah, the tributary king in Jehoiachin's place, and changed his name to Zedekiah, in commemoration of the oath then and there exacted from him in Jehovah's name, *to wit:* that he should give up the Egyptian alliance, and thenceforth remain a faithful vassal of Babylon.

PRELIMINARY OUTLINE. 185

This compact made, Nebuchadnezzar withdrew his armies, and returned to his capital, carrying with him the first group of prisoners recorded by Jeremiah, his own eighth year of reign ending with his departure, and it being still the center of 3406 A. M.

For a few years Zedekiah preserved his integrity, but under pressure of a strong party, who were conspiring to recover the Egyptian alliance, he at last broke his oath, and made overtures to Pharaoh Hophra Two years later Nebuchadnezzar's armies returned and came up to Jerusalem with the resolution to destroy it utterly.

From this point the dates of Ezekiel's prophecies accompany the events at Jerusalem, but, having already quoted them in a previous section of this Chronology, we shall now follow more particularly those of Jeremiah, who, as the special "Prophet of the Nations," and as one not yet vindicated—as to the fulfillment of the most important part of his mission—it behooves us to justify, and whose Chronology we shall set in order.

Space will not permit us to quote the texts referred to, but the running commentary will explain their gist, while it is of course incumbent upon the student to satisfy himself, by referring to the original.

" Thou shalt go to all that I shall send thee, and whatsoever I command thee thou shalt speak.

Be not afraid of their faces; for I am with thee to deliver thee, saith the Lord."—Jer. i. 7-8.

CHRONOLOGICAL ARRANGEMENT

OF THE LATTER PART OF JEREMIAH'S LIFE AS RECORDED IN THE BIBLE.

3415 A. M.   581 B. C.

XIVth year of Hebrew cycle, 10th month (4th civil), 10th day, Sunday.

The armies of Nebuchadnezzar arrive and besiege Jerusalem. Jer. xxxix. 1; lii. 4. (Comp. Ez. xxiv. 1-2.)

Jeremiah sent by the Lord with a message of advice and comfort to Zedekiah. Jer. xxxiv. 1-7.

Who thereupon causes a covenant of freedom to be made (perhaps at Passover?) Jer. xxxiv. 8-10.

Pharaoh's army now comes to the rescue (Jer. xxxvii. 5), and

The Egyptians capture Gaza. Jer. xlvii. 1-7.

Whereupon the Babylonians raise the siege of Jerusalem and depart to meet them. Jer. xxxvii. 5.

Zedekiah sends a petition to the prophet. Jer. xxxvii. 3-4.

Whose reply is dictated by the Holy Spirit. Jer. xxxvii. 6-10.

But Zedekiah and his princes break their covenant. Jer. xxxiv. 11.

And Jeremiah denounces them. Jer. xxxiv. 12-22.

The Prophet now attempts to go into the land of Benjamin. Jer. xxxvii. 11-12.

But is seized, falsely accused, brought before the princes, chastised, and cast into a prison which was in the house of Jonathan the Scribe. Jer. xxxvii. 13-15.

This incarceration seems to have been unusually severe. He was evidently treated as a rank political offender and traitor, and placed in the cells of the dungeon, where he remained "many days." Jer. xxxvii. 16. This is noticeably a reference to the days of "epact" at the end of the civil year 3415 A. M. (which was not an intercalary one, but nevertheless the 22 days then due, to rectify the Calendar, backed up into this year, and were a part of the XVth year of the cycle. Taken broadly the expression has its superficial English meaning, for Jeremiah was a prisoner until released by the Babylonians !)

3416 A. M.   580 B. C.

At length, or at the beginning of the new civil year, Zedekiah sent and took him out secretly, and having asked an important question elicited an equally momentous reply. Jer. xxxvii. 17.

Jeremiah now requested the king not to send him back to Jonathan's house. Zedekiah acquiesced, and directed him to be committed into the court of the prison. Jer. xxxvii. 8-21.

While there, in partial freedom, news of Pharoah's retreat and of the return of Nebuchadnezzar's army towards Jerusalem was soon followed by the actual fact, and the renewal of the siege early in the year. This induced Zedekiah to send an official commission to Jeremiah, who returned a message which seems to have been very generally overheard by all the people who were in the vicinity, and the news of which spread everywhere. Jer. xxi.

His particular enemies hearing that he had thus advised the people were wroth beyond measure. Jer. xxxviii. 1-3.

And now openly sought his life. At their continued insistence Zedekiah at last weakly gave the Prophet into their hands. Jer. xxxviii. 4-5.

So the princes took Jeremiah and cast him into the dungeon of Malchiah, which was also in the court of the prison. Here "Jeremiah sank into the mire." Jer. xxxviii. 6.

Ebed Melech now appealed to Zedekiah, and by his permission rescued the Prophet, and he was restored to his former quarantine in the court of the prison. Jer. xxxviii. 7-13.

After which Jehovah sent a message of safety to Ebed Melech. Jer. xxxix. 15-18.

It now appears that the princes were conspiring to have Jeremiah re-committed to the house of Jonathan, and had appealed to Zedekiah for permission. They intended to dispose of him. Jer. xxxviii. 26.

In the meantime Zedekiah sent for Jeremiah and held an important and final interview, in the principal entry of the Temple. Jer. xxxviii. 14-26.

After which the Prophet was re-committed to the court of the prison. Jer. xxxviii. 28.

The princes, attempting to investigate this interview, were misled, and thereafter were too much occupied with other matters, now at a crisis, to concern themselves about the incarcerated Prophet. Jer. xxxviii. 27.

In this final period of quiet, which spans the closing days of Zedekiah's reign (part of 11th year), an important word came to Jeremiah from the Lord. It led to a notable transaction, with an equally prophetic significance. This was his purchase, as a *"Goël"* of Anathoth, and the conveyance of the deeds, sealed and unsealed, to Baruch, for burial in an earthen vessel, which will yet be found in the land of the "Goëls," even in "Meath," where they speak with "Goëlic" lips! Jer. xxxii.

And yet again as the day of doom drew on, no doubt in the early part of Zedekiah's 11th year, a second, final, and superlatively significant prophecy, concerning the impending captivity and eventual return, *and a thrice repeated* GUARANTEE *as to the perpetuity of David's Throne*, came straight from God. Jer. xxxiii.

That prophecy meant naught, or it means all that Anglo-Israelites maintain it does, as to "Our Race" and to its line of Monarchs.

And now the time of Judah's probation ran fully out. The city fell (3416 A. M., 4th month, 9th day) and the Babylonians entered it. Jer. xxxix. 2-3; lii. 5-6.

Thereupon Zedekiah and his army fled *that night* (4th month, 10th day beginning) attempting a sortie through the Babylonian's lines. Jer. xxxix. 4; lii. 7.

But were pursued and captured near Jericho, and eventually brought to Nebuchadnezzar, who was himself far away in the North, at Riblah, in the land of Hamath. Jer. xxxix. 5; lii. 8-9.

Here he, Nebuchadnezzar, gave such direful judgment upon him, Zedekiah, that the very throne of David seemed to have been *shattered!* Jer. xxxix. 6-7; lii. 10-11.

[Into this error fell even Josephus, and since then, in it have abided all the Doctors of the Hebrew and the Christian Churches, since none of them have seen, and fearlessly maintained that a Throne whose perpetuity is so repeatedly and unconditionally *guaranteed* by the living Spirit of Inspiration, *could* not possibly have been brought to naught at this time, no matter how strong the circumstantial evidence thereto may be!

"The gifts and callings of God are without revocation" (Rom. xi. 29), his counsels are eternal, his covenants secure. The God of Israel is not a man that he should lie (Num. xxiii. 19–23). Hence though all men be proved liars, and all the generations of men be convicted of want of faith, yet nevertheless let God be true, and his oath to David salted with endurance!]

In the meantime the Babylonians under Nebuzaradan remained and sacked the city. Jer. xxxix. 28.

And obedient to the *particular*, and *noticeable* orders of Nebuchadnezzar, the chief men of the Babylonian army inquired diligently for Jeremiah, who was, of course, still in the court of the prison. Jer. xxxviii. 28.

News of his situation having at length reached them, they sent and took him out. Jer. xxxix. 13-14 (so far as the word prison inclusive).

And, for the purpose of temporary safety, appear to have sent him bound with all the other prisoners to the general rendezvous at Ramah. Jer. xl. 1.

Here he was probably released from his fetters, and held to await the official action of Nebuzaradan. At this Ramah, all the captives were rapidly collected, and such as were adjudged not worth carrying away were placed under the charge of Gedaliah, whom Nebuzaradan located at Mizpah, and made the governor of the subjugated province. Jer. xl. 7.

A month later the destruction of the city was completed (the burning lasted from the 7th day to the 10th). Jer. xxxix. 8; lii. 12-14. (Josephus).

Nebuzaradan's preparations for departure were now made, and pursuant to Nebuchadnezzar's original command concerning Jeremiah. Jer. xl. 2-4.

The general sent for him, and after a conference gave him an unconditional release, under circumstances of special and peculiar favor. Jer. xxxix. 11-12.

But Jeremiah was eventually advised to report to Gedaliah, and was *dismissed with a* GIFT. Jer. xl. 5 (compare Jer. xxxix. 14 from after word "prison" to "home" inclusive).

This advice Jeremiah elected to follow, so he departed, and came to Mizpah. Jer. xl. 6.

Where he dwelt. Jer. xl. 6; xxxix. 14.

With Gedaliah and the people. Jer. xxxix. 10.

Whom the Babylonians left behind them. Jer. lii. 16.

Then Nebuzaradan left Palestine with his captives. Jer. xxxix. 9.

And *en route* came up to Riblah. Jer. lii. 15.

Where Nebuchadnezzar still held the headquarters of his operations against all Syria, and where certain of this final group of prisoners were also put to death by Nebuchadnezzar. Jer. lii. 17-23, 24-27.

As soon as the hosts of Nebuzaradan had departed, all the fugitives secreted in the region gathered to Gedaliah. Jer. xl. 7-12.

And among them came Ishmael, whom Johanan secretly exposed to Gedaliah as a traitor. Jer. xl. 13-15.

But to no purpose, and the civil year ended. Jer. xl. 16.

3417 A. M.   579 B. C.

The events which followed upon the opening of the *New* year, are now detailed in their natural sequence in the book of Jeremiah.   Jer. xli; xlii; xliii.

In the meanwhile it will be well to notice the verse with which this straight account opens. "Now it came to pass in the 7th month that Ishmael * * * came to Gedaliah * * * and they did eat bread together in Mizpah." Jer. xli. i.

This was the civil *New year feast*. The sacred year, or calendar, always begins in March, and once in three years coincides, approximately, with the Vernal Equinox. Hence the seventh month here alluded to, being sacred, corresponds to the 1st civil month, or the month after the *Autumnal* Equinox. This too, is fully borne out by the fact that the fruits had ripened (Jer. xl. 10) had by this time been garnered and winter was approaching. Jer. xl. 12; xli. 8.

It was thus at a new year feast that the traitor Ishmael, slew Gedaliah; and, by seizing the KING'S DAUGHTERS soon after, attempted to support his own *pretentions to the succession*, by an alliance which would have *guaranteed them!*

It is also to be noticed that this particular year was the opening one of the new cycle, and the intercalary month had just been scored off upon both the sacred and civil calendars of (3416), in order to bring them into harmony, and to make their first day a first day of the week, as well as the first day of the first month after the Autumnal Equinox.

It was a high feast with much to be thankful for up to Ishmael's act of murderous treason and usurpation.

Such chronology as this is beyond the "higher criticism" (Selah!) and can no more be shaken by "job" and "hobby" chronologists than the planets can be shaken from their orbits. It not only scorns the leaven of the Pharisees, but defies the closest efforts of astronomy to prove it at all in error. And finally it stamps the Record as true history, with the indubitable seal of Divine authority.

The xli., xlii., and xliii. chapters of Jeremiah detail the incidents that followed this act of treason, and the close of their recital finds the Jewish colony settling in Egypt.

"All the people of the land" seem to have joined in this exodus, for when, a few years later, Nebuzaradan returned to the land to punish it for not sending tribute, and to wreak vengeance upon the Ammonites for aiding and abetting Ishmael, he found the land so empty that by scouring it he could secure but 745 Jews (Jer. lii. 30).

The very first act of Jeremiah, on arriving at Tahpanhes, where his own little remnant, consisting of Baruch, Ebed Melech, ZEDEKIAH'S DAUGHTERS, and a few trusty adherents, were placed by Hophra's kindness, was to bury the "great stones" beneath the clay of the brickkiln as a "sign" that Nebuchadnezzar should spread his pavilion over them, and set his throne upon them. The rest of the Jewish colony settled at Daphne, near by, where they gradually fell into the worship of Astarte.

In the meantime the siege of Tyre went on. Hophra's diversion against Nebuchadnezzar, by sea, reaped its reward, and reduced that of Nebuchadnezzar, and at length Pharaoh's Lybian expedition started upon its disastrous undertaking.

3428 A. M. 567 B. C.

At last the 13th year of Tyrian obstinacy came, and the proud city by the sea fell beneath the battering rams of the Babylonians. It was at this juncture that Jeremiah went to the feast at Daphne, and having pronounced Jehovah's displeasure, and the certain doom of all the Astarte worshippers, predicted the subjugation of Egypt to Babylon, and, as a "sign," announced the impending fall of Hophra (Jer. xliv).

Of course he was ridiculed; but, confident of his own mission and inspiration, he returned at once to Tahpanhes, and prepared for his own secret exodus, for he must have known that his original commission had its better part in store for him, and was at last about to materialize.

The Egyptian sky at this time was unclouded, and the doomed Pharaoh, although not without powerful enemies, was at the height of his glory. No one in that land, not even the most radical of the hostile political faction, could have anticipated what was even then, at the very moment of Jeremiah's utterance, transpiring in foreign lands.

At last, and only at the rate at which news could travel in that early day, the fact of Nebuchadnezzar's victory at Tyre, which portended the immediate transfer of his long delayed attention to Egypt, and of the utter failure of Hophra's Lybian expedition, arrived together.

3430 A. M. 566 B. C.

In the quickly succeeding confusion, incident upon the actual arrival of the Babylonians, and the internecine rebellion of Ahmes, the Book of Jeremiah ends, and the prophet himself and his chosen and favored remnant DISAPPEAR!

But so, too, Jonah disappeared,—from those who cast him overboard,—yet none the less Jehovah found conveyance for him, which, though strange, enabled him to prosecute his mission, and perform the will of God.

Jeremiah's disappearance from Eastern and Sacred History is the very reason why we should look for him in the SECULAR HISTORY OF THE

WEST. And in this western history we find him touching first at Spain,

3434 A. M.   562 B. C.

And resting finally in the westernmost of the Western Isles.

There he fulfilled his "mission;" there he "BUILT" and "PLANTED;" there, too, his dust awaits its resurrection, in the Isle of Davenish.

5889 A. M.
Monday, September 22d, 1890, A. D.

And there, finally, honor, and renown, and worthy reputation, await the successful spades of such archæologists as shall turn our tardy attention to the neglected tomb of Teah Tephi, and demonstrate unto the world whose Harp it was that was touched by David's Daughter within the halls of Tara!

Would that for every dollar spent in digging bricks out of the ruins of Babylon a dime might be forthcoming to explore this western mine of untold wealth, for if an indirect corroboration of the Scriptures be so valuable, in days of doubtful faith, what shall we say to such a lead as bids fair to reveal the very ARK OF ISRAEL!

The mystery of God's Romance is well nigh consummated. "The Tender Twig" planted by the Lost Prophet has spread abroad, until to-day the whole earth dwells beneath its far expanded growth. The "Riddle" of Ezekiel (chap. xvii.) has yielded up its secret, and the Anglo Saxon Race—"Our Race"— is about to stand forth in its final role as indeed JESHURUN, "a people saved by the Lord." It is thus with peculiar interest, in these hurrying days of the transition, that we are able to glance backward to the closing days of that prophet who was chiefly concerned in the transplanting and set their calendar in order.

THE
"END OF THE AGE,"
OR THE

LAST CHAPTER IN THE CHRONOLOGY

OF THE

"TIMES OF THE GENTILES."

"*And he said, Go thy way Daniel: for the words are closed up and sealed* TILL THE TIME OF THE END."—*Dan. xii. 9.*

"*For the vision is yet for an appointed time, but at the end it shall speak, and not lie: Although it tarry, wait for it;* BECAUSE IT WILL SURELY COME, — IT WILL NOT TARRY."—*Hab. ii. 3.*

"THE END OF THE AGE."

It will have been manifest to the candid reader of the foregoing volume that the object of the writer is two-fold. 1st. He has been actuated by a desire to vindicate to the human judgment the reasonableness of the Biblical Chronology, and thus to persuade it, if so be it is possible, to investigate this wondrous volume with more of the spirit of faith and veneration, and less of that which dominates this age of doubt. 2nd. The chief object has been to advance upon the irrefragible foundation of an astro-historical chronology, from which there is no escape, the far more momentous conclusions and warnings it has for the day and generation in which we live.

Between the Autumnal Equinox, which marks the publication of this volume, and the one which marked the "Mosaic Creation," detailed in Genesis i., and tabulated in the accompanying table, there extend exactly 5889 Solar years, reckoned from Equinox to Equinox ; and all the dates given in this volume belong to the one and only sequence of days and cycles which stretches between them and looks on beyond to time yet future.

In succeeding Studies of this Series, we hope to add numerous chronological tables to those in this one, and particularly do we desire to carry the main line itself, by means of the principal Biblical dates, from Creation down to the close of the Scriptural Canon, and from thence, by means of authenticated secular history, and the astronomical cycles, down to our present day.

In this undertaking we solicit the support and patronage of earnest men, and promise in return that the finished tables will shed an entirely new light upon the Scriptures. This effort is undertaken in the spirit of what in "Study Number One" we have already defined as "the *Highest* criticism," in contradistinction to its halting predecessor — the so-called "higher criticism." With the latter we have no sympathy—it has now become historically objectionable, its animus is evil, its library is apologetic, and its teachers should be rigidly catechised upon their Articles of Faith (1 John iv. 1–3).

We have fully set forth the superlative accuracy with which "Moses and the Prophets" wrote, and the scheme upon which the whole Biblical record is laid down.

We have purposely concentrated our studies upon its most momentous period—"The Babylonian Era"—since it is the earliest absolute origin of the "Times of the Gentiles," now apparently drawing to their close, WITH ALL THAT THIS IMPORTS!

It now remains to draw up a scheme which shall concisely show the *hither* end of the Scale,

CHRONOLOGY.
THE BIRTHDAY OF TIME.
THE HISTORICAL AND SCIENTIFIC STARTING POINT.

Year 1 Astronomical, or 0 A. M. 1656 before the Flood, 2555¼ before Joshua's Long Day.	Date of Month.	5889 Solar Years before the Autumnal Equinox of 1890 A. D.
		1st Civil Month, Autumnal Equinox.

First Week {	First Day, ★ Sunday.	1	I. The cycle of the Hours begins.
	Second Day, Monday.	2	II. The Solar year begins with the first day of the week (apply Solar cycle as a test).
	Third Day, Tuesday.	3	
	Fourth Day, Wednesday.	4	III. The ancient Solar cycle begins, seven Lunar years. (Antediluvian) intercalated.
	Fifth Day, Thursday.	5	
	Sixth Day, Friday.	6	IV. Common Team of Eclipses begins.
	Seventh Day, Saturday.	7	V. The Lunar cycle begins, *i.e.*, Metonic.
Second Week. . . . . {		8	VI. A Moon begins agreeing with ours.
		9	VII. Scriptural History begins. No dates fail to accord with this "line of time."
		10	
		11	
		12	VIII. The maximum cycle of Eclipses begins.
		13	
Second Sabbath. . . .		14	IX. Proved by the transits of Venus.
			X. Proved by the transits of Mercury.
Third Week. . . . . {		15	XI. Prophetic times and cycles commence.
		16	
		17	XII. The week begins, agreeing with present sequence.
		18	XIII. All the rectified dates of secular history corroborate this date.
		19	
		20	
Third Sabbath. . .		21	XIV. The Equinoxes agree thereto.
			XV. The genealogies of the Bible agree.
Fourth Week. . {		22	XVI. Finally, all astronomy and history that does not agree thereto is necessarily *bogus*. Beyond it there is nothing "*prehistoric*"— geology, evolution, and disbelief to the contrary notwithstanding.
		23	
		24	
		25	
		26	
		27	
Fourth Sabbath. . .		28	"In the beginning," of which Moses wrote the concentrated activities of "Elohim" laid the strata of the earth as set forth in Genesis, and we have no *ex post facto* basis upon which to judge the results. Chronology corroborates the account. This is sufficient.
Part of Fifth Week. . {		29	
		30	

and to synchronize the several subordinate sets of "Times" against it so that the warning which it utters may be read by "one that runs."

If the point from which we start the scale (3377 A. M.) cannot be shaken,—and the author is satisfied that it cannot, so long as the Solar System preserves its astronomical integrity ; and if the scale to be employed is that of the solar years (for the lunar years ran out in 1824 A. D.), —then the point *to which* it is progressing is the CRISIS of the Human Race.

Upon far less evidence than what he has set forth in this volume the writer would not hesitate to become a Jonah to a Nineveh ; certainly he can be fairly accredited with honesty of purpose, and concern in premises that seem to affect the future not only of his own "Race" but of all mankind as well !

With these few final words he therefore submits the accompanying modern terminus of the scale, which, measured by the same units, comprehends the "Seven Times" of Gentile Rule, and synchronizes its last seven years with the common A. D. Calendar. It is a practical diagram, and compasses "THE END OF THE AGE:" it stands in natural apposition to the STARTING-POST OF "TIME" above given, and between these two termini all the history of OUR RACE, from Eden to the "consummation of all things," must find appropriate place.

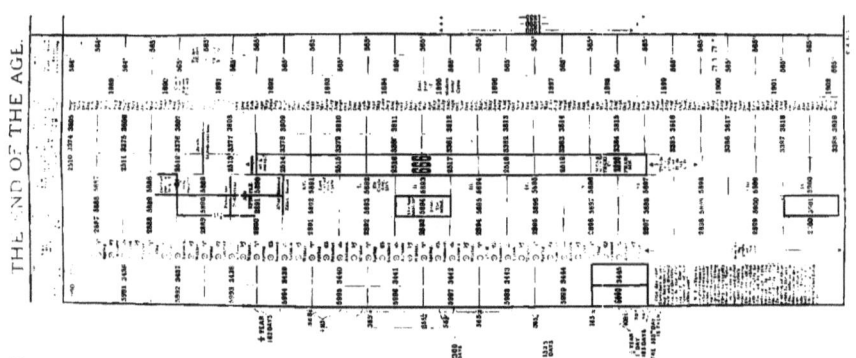

As a student of chronology of many years' standing, and as one quite familiar with other systems, he does not hesitate to pronounce the one herein followed to be the only correct system.

Not because it is his own, for this he disavows. It is God's own, so near as human study has yet compassed it, and it is the logical outcome of a Human Science which has not *stultified* itself with "Infidelity."

The times are now short and their Signs are all completed save a single one—the manifestation of "Ho Anomos" "That Lawless One" (2d Thess. ii. 8), whose synonym in the same language gives us the familiar neologism, "Ho Anarchos"— (THE ANARCHIST)—and these short days (a year and a half) are the SOLE DAYS OF GRACE THAT YET REMAIN TO US. For when that One, "The Mystery of Iniquity," in its final phase shall have begun his reign, the Holy Spirit, which hitherto has withstood it (2 Thess. ii. 6), will have withdrawn! (7.)

From that dread moment onward we must date the "Great Tribulation" which is the Time of "the Harvest."

It is therefore with deep concern that the author submits the accompanying table, which so inevitably results from the dread premises he has been able to establish, and leaves to those who will consider it, and to whom this volume is dedicated,

As a student of chronology of many years' standing, and as one quite familiar with other systems, he does not hesitate to pronounce the one herein followed to be the only correct system.

Not because it is his own, for this he disavows. It is God's own, so near as human study has yet compassed it,.and it is the logical outcome of a Human Science which has not *stultified* itself with "Infidelity."

The times are now short and their Signs are all completed save a single one—the manifestation of "Ho Anomos" "That Lawless One" (2d Thess. ii. 8), whose synonym in the same language gives us the familiar neologism, "Ho Anarchos"— (THE ANARCHIST)—and these short days (a year and a half) are the SOLE DAYS OF GRACE THAT YET REMAIN TO US. For when that One, "The Mystery of Iniquity," in its final phase shall have begun his reign, the Holy Spirit, which hitherto has withstood it (2 Thess. ii. 6), will have withdrawn! (7.)

From that dread moment onward we must date the "Great Tribulation" which is the Time of "the Harvest."

It is therefore with deep concern that the author submits the accompanying table, which so inevitably results from the dread premises he has been able to establish, and leaves to those who will consider it, and to whom this volume is dedicated, the task of drawing from it such warning as they may.

I am deeply indebted to another Yale professor for the true insight into this ii. chapter of 2 Thessalonians, which is hereinbefore set forth. This chapter is the veritable KEY to what is *now* working, and to what is soon to come to pass. That which "letteth" or resists, or hindereth the "Spirit of *Evil*" is the *Holy Spirit*, and *not* the Roman Empire!

"The Comforter" was sent to us after the Saviour's Ascension, to be with the Church of Christ until the end. Progressive interpretation of the Word now suggests the awful certainty that the Holy Spirit, grieved beyond endurance, will withdraw before the Second Advent!

With it "the Elect" will probably be "caught up," to join the returning Saviour in the air.

But "woe to the inhabiters of the earth" (Rev. xii. 12) when "the wise virgins" disappear! The "foolish" ones will then be truly surrounded by a pack of wolves, for when the HOLY SPIRIT "withdraws itself" man *must literally face the* INCARNATE DEVIL!

CONCLUSION.

THE GENERAL SKELETON
OF
"THE TIMES OF THE GENTILES."

A SOLI-LUNAR HARMONY,

2520 YEARS.

"*For this is the Day of the Lord God of Hosts, a Day of Vengeance, that he may avenge him of his adversaries; and the sword shall devour and it shall be satiate, and be made drunk with their blood; for the Lord God of Hosts hath a sacrifice in the north country by the river Euphrates.* * *

"*But fear thou not, O my servant Jacob, and be not dismayed, O Israel; for behold, I will save thee from afar off, and thy seed from the land of their captivity; and Jacob shall return and be at rest and at ease, and none shall make him afraid.*

"*Fear thou not O Jacob my servant, saith the Lord, for I am with thee, for I will make a full end of all the nations whither I have driven thee, but I will not make a full end of thee, but correct thee in measure; yet will I not leave thee wholly unpunished.*"   (*Jer. xlvi.* 10, 27–28.)

THE CONCLUSION OF THE MATTER.

We have now reviewed, upon a scientific basis, some of the chronological inferences to be drawn from the rectified "scale of time." In our final table we present a general survey of the chapter which is *now* of chief importance. Manifestly it behooves the "Church" to be prepared. It is idle to say, in the face of these facts, that the beginning of the end *does not* synchronize with the issue of this present volume; it is equally idle to maintain that it does. However misquoted, therefore, the present author may hereafter be, he wishes to place on record a fair statement of his position, to wit:

(*a*) In view of the Saviour's own command (Mark xiii. 33), it is our duty to be ready.

(*b*) In the spirit of Daniel ix. 2, it is equally our right and duty to *study* "by books the number of the years."

(*c*) In the same way as Daniel's efforts were at length rewarded (Dan. x. xi. xii.), we may hope for general light.

(*d*) In the meantime we are expressly told that no one need doubt his senses when the thing is nigh, even at the doors (Mark xiii. 28-30). There can be no doubt that this is so TO-DAY.

(*e*) And we are also assured that this dispensation (this "Israel," this Age of the "Goyim" or "Gentiles," this "Era of Evil," typified by Babylon, and dating from her correct origin, "shall not pass away till all these things be fulfilled" (Mark xiii. 30).

(*f*) Now the scheme of Chronology which we endorse, and which squares itself against history and astronomy, must remain unimpugned whatever happens, and if we have applied our prophetic scale to the correct *beginning*, then is it equally manifest that it reaches the proper *end*, and therefore must span all else that lies between us and that termination.

(*g*) If the Times of the Gentiles, of which Nebuchadnezzar *and his father*, quite as much as Nebuchadnezzar and his sons, stood for "the head of gold," began in 3377 A. M., they end in 5897 A. M. (or at our March, 1899); and it need not surprise the Church that such a truth should be revealed, in its proper season, since we are told (Hab. ii. 2-4) that "in the end" it shall be "plain" to one "that runs," the which Daniel also fully endorses (xii. 4-8, 9). Nor dare we hesitate to avow our positive conviction that the time is even now upon us.

(*h*) Yet let no weak vessel hereby be overturned, so that its quantum of faith be spilled, if, by the Grace of God, these days be still further lengthened, and so another soul be added to his sheaves. In other words, we must point out

CONCLUSION OF THE MATTER. 213

that it is *possible* (although from the consummate way the scales of Prophecy and History harmonize when adjusted from 3377 A.M., we must confess it is not at all *probable*), the commencement of the Solar Era may eventually be measured from Nebuchadnezzar's own second year, *i. e.*, from 3399 A. M. It was in this year that he dreamed the vision which was taken as the "type" (Dan. ii. 1), and it was unto him *personally* that Daniel said (36-38) "*Thou* art this head of gold."

(*i*) By no possible method of interpretation can we slide the scale of 7 Times or 2520 Prophetic years, down that of fixed chronology, beyond this final point, and to do so would increase our "margin of grace" by only 22 years.

(*j*) We have said that this "increment" is not probable. This is of course only our own judgment, but it is founded upon a collateral survey of numerous other prophecies, whose interpretation seem inevitably to focus upon the end of this century, and not to span over into the next.

(*k*) Nebuchadnezzar was, and is, "this head of Gold." In SOLAR time the 2520 years commence with the Chaldee Babylonian Era, and run out in 1899 A. D., but in their *personal* application to Nebuchadnezzar himself, they clearly delayed, until no longer able to resist his besetting sin of pride he boasted in his palace, and exclaimed:

"Is not this great Babylon which I have builded for the house of the kingdom, by the might of my power, and for the honor of my majesty?"

And, "While the word was in the king's mouth, there fell a voice from heaven, *saying*, O King Nebuchadnezzar, to thee it is spoken; The kingdom is departed from thee." Dan. iv. 30-31.

From this time the vision of the Image "speaks"! A consultation of the accompanying general diagram will demonstrate the fact: for seven years the king, as a type, was insane, for half a year he recovered his reason and acknowledged his sin in an epistle unto all nations. Then, his problem solved, he died; and the times betokened by the vision, and the type, began their shorter, LUNAR course. From 3444 A. M. to 5888 A. M. inclusive (2445 solar years, or exactly 2520 lunar years), the "times" repeat upon an *included* scale, and run out at the Autumnal Equinox of 1890 A. D.

They are ended! Just ended! And now a pause of a year, and of a half a year, come in between them and the final seven which are to antitypify all that heretofore has taken place.

These final seven, *the reign of* ANTI-CHRIST, consummately complete the scheme and make it SOLI-LUNAR to the last degree.

Stood we condemned to an inevitable death, upon one-thousandth part of the certainty involved in these calculations, we would all be engrossed in setting our affairs in order, and shall

THE CONCLUSION OF THE WHOLE MATTER
OR A
GENERAL SKELETON OF "THE GENTILE TIMES"
AS SET FORTH UPON THE
BIBLICAL OR SACRED "CALENDAR."

TRUE A. M. YEARS.

2520 Solar Years.—Dan. ii. vii. viii. ix. x. xi. xii.
{
    65 {
        21.y. { From middle of each, inclusive.
        3377½   "And *because* the Bridegroom was *long away* they all slumbered and slept."—Math. xxv.
        3398½   (The true translation as rendered by the best Greek scholars.)
        3399   Dan. ii.

        3433   Dan. iii.
        3434   Neb's Vision.—Dan. iv. 1-27.
    45 {   3435½   Dan. iv. 28-33.
        3436½
        3437½
        TYPE.   3438½   } 7 Nebuchadnezzar's "INSANITY."   N. B.—The central point of the whole era is **4636 A.M.**
        3439½
        3440½
        3441½
        3442½   Dan. iv. 28-33.
  ½ {   3443   Neb.'s Epistle to all Nations Written.
  1 {   3444   *Vide* Dan. iv. 34-37.

    2445 {
        2nd Thess. ii. 6,7-17.
        2520 LUNAR OR 2445 SOLAR YEARS inclusive
        (HABAKKUK) Chap. ii. 3.
        THE "SHORTENED" "SEVEN (LUNAR) TIMES" OF JUDAH, Mercifully Reduced.   **4666**=Central Year!
        (MARK Chap. xiii. 20.)   } The final 7 years must be float'd into modern A.D. years.
        5888
  1 {   Mal. iv.   5889   2 Thess. ii.   Zach. xiv.
  ½ {   5890½   2 Thess. ii.   1892¼
        5891½   Rev. xiii.   3¼
        5892½   "ANTI-CHRIST'S" Dan. ii.   4¼
  7 {   ANTI-TYPE.   5893½   } 7   REIGN   Mark xiii.   Rev. xiii. 18.   ★ 5¼
        5894½   OF   Dan. iii.   6¼
        5895½   HORROR.   Dan. xi.   7¼
        5896½   Rev. iv. xix.   "666"   1898¼
        5897½   To Middle of Year.   James v.   1899¼

1899 A. D.=6000 A. M.   Soli-Lunar=1335th year.—Dan. xii. 12.

Sometime, from now on, after the final 7 years have been fully begun, and "the Lawless one" has been identified (2 Thess. ii. 3, 8) Redemption draweth nigh. There is, therefore, now but one Christian motto, "And what I say unto you I say unto all, WATCH!"—St. Mark xiv. 32.

we not awake to an alarm which loudly proclaims that the DIES IRÆ is at hand?

(*l*) And finally; what is it to Thee, O Man, seeing that thou art enclosed in the net of this final generation anyway? As one of its "units" *thou* mayest pass away to-morrow (James iv. 13-14), and, if with unreplenished lamp thou goest, then, clearly, thine *own* equation will be as fully solved as if perchance the Lord's coming fell upon this very day, and oil were wanting in thy vessels.

It is to this end chiefly that we have supplemented our own earnest testimony in behalf of things so meet for the consideration of "Our Race," with seals of sufficient dignity and authority to demand respect among Scientists, and so-called "higher critics."

As it is easier to tear down than to build, let now all such as dispute the "Chronology of the Bible" as herein vindicated, produce their own case, and show wherein the spirit which inspired "the Record" has failed to adhere to its own set "times and seasons." But in the meantime let all such as are "wise" perceive the significance of this conclusion, to wit: that if "Joshua's Long Day," and the "Shadow upon the dial of Ahaz" are hereafter to go back into history as integral parts of the now fully vindicated record of "Our Race," then verily unto it do they also raise their voices and unite with all cycles of heaven in A MIDNIGHT CRY!

EDITORIALS.

"*And Joshua said unto the people, Sanctify yourselves: for* TO-MORROW *the Lord will do wonders among you!*" Joshua iii. 5.

ϴUR RAGE:

ITS ORIGIN AND ITS DESTINY.

SERIES I.   SEPTEMBER, 1890.   NO. 2.

EDITORIALS.

"The time will come when Bible Prophecy with its Chronology will be confirmed by History in so exact and so signal a manner that malice and infidelity alone will be able to deny its inspiration. Then, too, the world will have had its last say, impiety will have let fall its last masque, intolerance have practiced its last cruelties, superstition descended the lowest round of idolatry, faith won on the scaffold its most brilliant victories, and in presence of the last great revolution History will learn from Prophecy to comprehend and to judge itself. The transformation which it will experience will be so complete that a very small remnant will be found of what it to-day calls its Philosophy."

Thus wrote Frédéric de Rougemont, the earnest Swiss pastor, nearly a generation ago, and behold we are already standing upon the threshold of the days to which he alluded. The New Chronology has come to stay, and unto it whatsoever of History is worthy of survival must hereafter conform all its references. We dare to say this *ex cathedra*: for, satisfied that the Bible is inspired, even in the sense accepted by the wise of all former ages, and being ready to demonstrate that its hitherto most inexplicable cross references yield without violence to the requirements of this rectified chronology, we risk nothing, save perhaps a chapter or so of modern criticism, in stating our position plainly. That we shall incur a torrent of animadversion for our temerity we ought perhaps to have no doubt, we expect it from the quarters of the arena into which we have hurled the gauntlet, but we ask the audience to require all combatants to raise their visors ere they join the fray so that it may be known from the start upon which side they are truly crossing swords.

*

As the whole of this present Number of the Our Race Series partakes of the nature of Notes, Queries and Replies, and has a direct bearing upon the Israelitish origin of the Saxon Race, we omit the few pages which would otherwise have been devoted to them. In the meantime we take occasion to thank the numerous correspondents who have sent us data of interest touching the subjects advanced for discussion in Study Number One. If the replies and material continue to come in at the present rate we shall in time be able to devote an entire volume to them, and promise to present an array of facts and circumstantial evidence which it will be impossible to combat. We are satisfied, however, it will be agreed by all that Chronology is now the chief matter of concern.

It is the fundamental basis of all accurate historical investigation, and not until its skeleton has been satisfactorily articulated can we hope to clothe our topic with flesh and nerves, and vitalize it with the blood of life. Henceforth we shall give all dates in the true chronology—*i. e.*, the Biblical or A. M. years. Those given in Study Number One were upon the common A. D. and B. C. scale, and should be translated into the A. M. scale by means of the table given upon page 113.

**

They must be blind indeed who fail to read the warning written upon the walls of the modern Temple of Theology. In its continued subdivision into sects the Protestant Church has had its strength so decimated that, as the Master long ago predicted, it is doomed to fall (Matt. xii. 25). We Protestants are prone to draw invidious comparisons against Rome, while she in turn points out the ever widening breaches which divide our house against itself! Just where the balance of error actually resides is hard to tell. The fact is the spirit of Laodicea presides over the whole city of Modern

Babylon whatever be the particular ward in which we dwell, and the cry should now go up throughout all its precincts, "Come out of her my people, that ye be not partakers of her sins, and that ye receive not of her plagues" (Rev. xviii. 4). This is the Midnight Cry itself, and it appeals to all "the wise," wherever they are domiciled, to *go out* to meet their coming Lord, and to take naught with them but that oil which burns with the bright flame of faith in the integrity of the whole Bible. That we ourselves are dwelling in this Laodicean Babylon is patent to all who are familiar with the methods upon which its "*primaries*" are conducted, nor can we fortify our assertion better than by quoting once more from our trenchant Swiss pastor, who wrote as follows of a state of the Church, in his day future, but now, alas! only too realistic:—

"The closing Epistle of Christ to the Seven Churches (Rev. iii.) is directed to Laodicea. It corresponds to the times of Jewish phariseeism and sets forth the state of the Protestant nations at the Lord's return, when there will be little or no true faith left on the earth. The missionary zeal of the Church of Philadelphia, which at one time enflamed the whole mass of reformed Christianity, will have subsided into lukewarmness. The whole area will be Christian, and pride itself on its profession. A high standard of morality, an upright life, a conservative creed, will be never so *popular*. There will be no open enemy of

Christ as in Philadelphia, no outspoken infidel; only phariseeism and lukewarmness, only the happy medium between impiety and pietism. There will be a little faith, but not too much; a profession of orthodox principles, confined within wise limits. There will be some fear of God, but much fear of men; great respect for the Bible, but enough good sense to keep men from viewing its doctrines, its precepts and its denunciations in a serious light; society wholly given to the acquisition of temporal blessings, and yet diligent enough in public worship not to doubt the pardoning mercy of God. They will consider themselves very rich in Spiritual life, they will even have need of nothing. But the Lord will *vomit lukewarm Laodicea from His mouth.* He will not fight against her, as against Pergamos and Thyatira, He will not judge her like Sardis; but He will wholly cast her off with scorn, and leave her to her wallowing in the mire. Still, she is a church, and Oh, mystery of grace! He even speaks to her of *love.* He counsels her, rebukes her, treats her like a child subjected to salutary discipline: 'I would that thou wert cold *or* hot.' 'Be zealous, therefore, and repent.' He offers her a *collyrium,* that she may open her eyes to her *wretched state;* the white raiment of His righteousness, that the shame of her nakedness may not appear; gold tried in the fire of faith, that she may be truly rich. But His offers will not be accepted by the vast majority of the

Laodiceans; few of them will even hear His voice when He 'stands at the door and knocks' to invite His guests to the bridal supper. Those, however, who in the midst of the universal apathy have persevered in love to the end, will receive the highest honor of all the faithful: they shall sit down with Jesus on His throne.

"The Church of Laodicea is no far-fetched type; it mirrors the Protestant world to-day, and its distinct presence is not one of the least of the sign-posts that guide the weary pilgrim along his midnight highway."

***

If a tithe of what we have reviewed in this present Study is even approximately right, then certainly this generation is to witness the harvesting of God's vineyard, and if, as we believe, it is wholly true, then must *this very decade see the scythe thrust in!*

"Now Jordan overfloweth all his banks all the time of harvest," and so, too, doth the river of Prophetic fulfillment inundate the banks whereon those who bear the Ark of Truth are dipping the soles of their feet into "the brim of the water" (Josh. iii. 15).

But no matter how its waters swell Israel is to pass over dry shod—the waters must roll back even to the City of Adam, and below us they must fail from their channels, even to the Salt Sea. We must "pass over right against Jericho!" There are stirring times ahead of us, and ere we may count upon a *new* "division of the land," a great and terrible DAY is certainly in store for all the earth, a day when those who doubt the truth of Joshua's Long Day *are doomed to see its very opposite take place!* for:

"It shall come to pass in that day saith the Lord God, that I will cause the sun to go down at *noon*, and I will darken the earth in the clear day" (Amos viii. 9).

* * *

"And he said unto me, Seal not the sayings of the prophecy of this book, for the time is at hand.

He that is unjust let him be unjust still: and he that is filthy let him be filthy still: and he that is righteous let him be righteous still: and he that is holy let him be holy still.

And behold I come quickly: and my reward is with me, to give every man according as his work shall be" (Rev. xxii. 10-12).

"*Woe for the earth and for the sea: because the devil is gone down to you having great wrath, because he knoweth that he hath but a short time.*" *Rev.* xii. 12.

MISCELLANEOUS.

A CARD.

Had we the means, we would willingly give a copy of these works to every human being, but while this current dispensation lasts, we are unfortunately forced to "sell the truth," (Matt. xxv. 1-3), to those who know its present value. (Prov. xxiii: 23)!

THE OUR RACE PUBLISHING CO.

OF THE

OUR RACE, ITS ORIGIN, ITS DESTINY, SERIES,

WE WISH TO CALL ATTENTION TO

STUDY NUMBER ONE,

ENTITLED

THE ROMANCE OF HISTORY:
"LOST ISRAEL FOUND."

By Prof. C. A. L. TOTTEN, (Yale Univ.): with Introduction by Prof. C. PIAZZI SMYTH, (late Astron. Royal, Scot.)

A Unique 12mo; 288 Pages; PRICE 75 CENTS.

Antique Binding.

This is pre-eminently **THE** volume of the **HOUR** which is striking upon the dial of the **AGES**. It treats of the Emergency questions which now lie at the Anglo-Saxon Door, and its clarion summons should arouse our "Royal Race" from apathy and sleep, and accelerate the consummation of its *Mission*. With significant arithmography the author has concentrated the destiny of this dominant people into an acrostic composed of the vowels of their universal language,

A. E. I. O. U. Y.

ANGLIAE EST IMPERARE ORBI UNIVERSO YISRAELAE.

It is for the Anglo-Israelites to dominate the Universe!

Like the Race, of whose history this volume treats, the book itself has a *past*, a PRESENT, and a FUTURE, and we want earnest agents to put it into earnest hands. The first edition, a *limited* one, is being rapidly exhausted, and almost every volume called for seeds down an immediate demand for numerous others. All who have read "Our Country," by Josiah Strong, should make haste to secure this still more comprehensive survey of *our* Origin and Destiny. They will save time and insure personal attention by ordering it directly from the publishers. Our Company has been *incorporated* under the laws of Connecticut for the express purpose of spreading the TRUTH broached in this opening volume; the unusual incidents leading up to this step are fully set forth in the book itself; they will be a revelation to many!

The volume is popularly written, and its rhythm is *in touch* not only with its own *motif*, but with the *Zeit-Geist* or "spirit of the times." From among the commendations of the few to whom its "Advance Sheets" were submitted we select the following :

"It is so new, so strange, so startling."—**Joseph P. Bradley** (Justice U. S. Supreme Court). "But little short of inspiration."—**Rev. Emerson Jessup.** "I would not have believed that you could have put me—a country outsider given to chopping and literary *excursus*—into such quick and lively *rapport* with the issues you discuss. Your enthusiasm is catching, and I am sure must catch readers in abundance."—**Donald G. Mitchell** (Ik Marvel). "The most readable book for the general public yet published."—**Rev. Geo. W. Greenwood** (late Editor of The Heir of the World). "Will be widely read."—**Hon. Edward J. Phelps** (Ex-Minister to Great Britain). "I have learned sufficient to make me ponder and search."—**Rabbi A. P. Mendes** (Touro Inst., Newport, R. I., "Nobly written and scripturally founded."—**Prof. C. Piazzi Smyth.** "Just the thing needed."—**Edward Hine.** " Your theme is a noble one, and one which ought to engage our reverend, careful, humble, long study. If the case can be fairly made out, nothing so noble has crowned all the Scientific, Historic or Scriptural research of these wonderful days of ours. It would (as does the presence of the Jews as a distinct Race, and far more, I think, than that) afford a wonderful confirmation of the Sacred Writings. It would be a proof before our very eyes."— **W. W. Niles** (Bishop of New Hampshire). " When your books are ready I shall try to spread about a score of them; in the meantime please find $25 to render a little help."—**J. W.** (This is but one of many letters of a similar generous nature, and in an age whose mercenary motto is that "Money talks" speaks with emphasis!) "I will take One Hundred dollars' worth of the books ; I do not wish them sent to me ; I will go for them myself, and I shall scatter them in every direction."—**C. A. G. L.** —— "I am fascinated with the 'Romance of History.' In my opinion God is using you to make plain one of His grandest objects in creation."—**Chas. W. Carpenter.** "I am on the second reading of your book, and it impresses me more strongly than it did at first."—**Thomas Ridgway** (U. S. Army).

Such testimonials continue to pour in, now that the volume has begun its pilgrimage, and we are convinced that they are simply the " wave sheaves " of a tremendous harvest Help us to reap it, for we need laborers in the vineyard. The topic is one that comes home to every Anglo-Saxon, and at this juncture, in a special way to every patriotic American, who hereafter may truly say—"I too am of Arcadia."

Send price (check, money order, or postal note), with your address to

The Our Race Publishing Company,

P. O. BOX 1333, NEW HAVEN, CONN.

WORKS BY THE SAME AUTHOR.

YALE MILITARY LECTURES. Selected from Series of 1890. First Section: National and International. I. (Introductory) Lecture—The Military Outlook at Home and Abroad. II. Military Economy, and the Policy of America. III. The Military Problem of America, with Notes on Seacoast Defence. IV. Organization, Disorganization, Reorganization, Mobilization. 1 Vol. With Illustrations and Tables. 1890. Price 50 cts. Send orders to Editor of " Our Race," P. O. Box 1333, New Haven, Ct.

The extra-large editions of the *New Haven Register*, in which these lectures originally appeared, having been so quickly exhausted, they are now reproduced in convenient book form. This is done in order to satisfy the continued demand for them, due, no doubt, not only to the novel treatment of the topic of the Second Lecture, but probably more particularly to the Prophetico-Historical exegesis of "The Signs of the Times" contained in the First. The whole series is written in the spirit of *Anglo-Saxon Identity with Israel*, and the response from all quarters has shown that the topic is "in touch" with a hitherto deeply latent, but none the less real, American sentiment.

"These lectures are valuable historically and economically. They deal with a vast subject, which is of the highest importance to the future welfare of this nation. They are written in a popular vein, and are thus brought within the easy understanding of all classes of readers, particularly those interested in the political and social questions which concern our progress. We recommend these lectures to the people."—EDITORIAL, *New Haven Register*, Jan. 13, 1890.

"The treatment of the subject of your second lecture is original, and as forcible as it is comprehensive. It is addressed to a larger and more mature audience than those usually found in the class-room; and if the whole course be pitched upon the same key, it will be well worthy of publication in permanent form for general circulation."—Frank G. Smith (Capt. and Bvt. Maj., 4th U. S. Arty.).

"I feel that you are doing a good work, not only for the students, but also for the general public."—H. B. Bigelow (Ex-Gov. Conn.).

"It is no new thing to find military men interesting themselves in studies and speculations of this nature,—witness the case of the late Gordon Pasha,—and so there is no real occasion for surprise in the circumstance that Lt. Totten . . . is combining with his more commonplace and matter-of-fact function that of an expounder of the prophetical writings. He looks to see the existing governments of Europe give place to democracies, which will speedily run into atheistical anarchies (such as the Paris Commune gave us a glimpse of nineteen years ago), and fill the Old World with bloodshed, renewing on a vaster scale, and surpassing, the butcheries of the French Terror. One of the lessons deduced by the lieutenant is an eminently practical and professional one. He would have this country fortify its coasts and strengthen its navy betimes, that, when that lurid storm bursts upon the earth, it be not taken unawares and at disadvantage."— Editor *Hartford Courant*, February 11, 1890.

STRATEGOS. To which is appended a collection of studies upon Military Statistics as applied to war on Field or Map. 2 vols. Illustrated. D. Appleton & Co. 1880. Price, $3.00.

"A careful consideration of the *statistical* merits alone of this work will recommend the new line of investigation proposed therein as worthy of the diligent study of all concerned."— Alex. Ramsey, Sec. of War.

"After a thoughtful perusal of its contents, I can only add that this very interesting publication, based upon the most careful considerations, warrants the possibility of any one following these studies either alone or in company with comrades, from the very simplest tactical evolutions, combinations, manœuvres, and battle plans, in systematic gradations up to military operations, and also of using the same for the prosecution of the history of the later wars."—G. Von Moltke (General Field Marshal).

"It will do much to impart military knowledge and the science of Strategy to many who without it would never have turned their attention in that direction."—GARNET J. WOLSELEY.

"Concerning your method of Kriegspiel I take pleasure in testifying to the praiseworthy distinctness and excellent systematic order of the material. It contains so many new and practical hints for us, that it was very highly recommended already for general study in the Swiss Military Journal."—H. BOLLINGER (Colonel, Swiss Mil. Academy, Zürich).

AN IMPORTANT QUESTION. A study of the *Sacred Cubit* of the Hebrews, as the undoubted *origin* of ANGLO-SAXON measures. 1 vol. Illustrated. John Wiley & Sons, 53 East 10th St. N.Y. 1887. Price, $2.50.

"The more I read of Lieut. Totten's writings the more I respect his learning, his ability, his mathematics, his chivalry in the cause, and his religion. I am abstracting just now from his book (Important Question) into mine, so that I may recommend readers to purchase his, and am letting them know where to write for it."—C. PIAZZI SMYTH Ast. Roy., Scotland.

"From the scientific standpoint this volume must receive wide attention. There is something so new and startling in its method of treating physical data, that it seems as though an entire scientific method had been discovered at once. The volume is a bold challenge to President Barnard and the advocates of the metric system to produce their case and put it upon the same or an equal basis."—*Army and Navy Journal.*

"After perusing such a volume one can readily comprehend the words of wisdom (xi. 20): 'Thou hast ordered all things in *measure, number,* and *weight.*' It is out of the question to review such a work, or to give any consecutive idea of its contents. It is one that every Anglo-Saxon should study for himself."—*The Evangelist (N. Y.).*

"If the facts and possibilities suggested by Lt. Totten in this connection are as stated, there can be no doubt of the superiority of our ancient and time-honored system over the one which is striving to supplant it."—*N. Y. Herald.*

"It contains new and startling scientific facts evolved in a most unexpected way from old and familiar things."—*N. Y. Mail and Express.*

"The appendix upon the 'Sacred Cubit' is an extraordinary study in geometry and algebra, while the volume as a whole is a monument of special learning."—*N. Y. Times.*

"His system of metrology is cosmical; to call it ingenious would be a tame meed of praise. It is the development of a genius."—*International Standard.*

THE FACTS AND FANCIES, LEGENDS AND LORE, OF NATIVITY. Illuminated by Tiffany & Co. Oblong 8vo, cloth, gilt edges, bevelled boards. Price, $5.00. An elegant giftbook. John Wiley & Sons, 53 East 10th St. N.Y. 1887.

"This is one of the most unique and interesting volumes that ever came from the American press. . . . It is a mosaic of original ideas extending over the whole range of legend and literature, filled with facts and with quaint and curious lore. . . It is no ordinary birthday book, it is an ideal book of Nativities challenging the curiosity of the curious, and furnishing to lovers of gem-lore and sentiment an inexhaustible mine of suggestion, information, and enjoyment."—Extract from review in *Education,* March 8, 1888.

INSTRUCTIONS IN GUARD DUTY. Complete, and for use upon the spot. Prepared for the C. N. Guard. 1887. Limited, Vest-Pocket Edition. Price 25 cts. Address the Author, 77 Mansfield Street, New Haven, Ct.

This is literally a vest-pocket companion, covers the whole subject in a nut-shell, is accurate, brief, pertinent, and in conformity with the customs of the service. It is just what the National Guardsman needs. It scans every duty, of every grade, in concise notes, headlines, and practical reminders, and absolutely suits the circumstances for which it was intended.

"It would be a good investment for the Adj.-Gen'l to supply the armories with a few copies of this little work, as it gets at the subject *quickly* and *closely* "—C. R. DENNIS (Q. M. G. R. I. M.).

"Guard duty showed marked improvement at the last encampment, and with careful study of the valuable manual 'Instructions in Guard Duty,' by Lieut. Totten, U. S. A., a number of copies of which will be shortly issued to each company, still greater improvement should be shown at the next encampment."—FRED'K E. CAMP (Adj. Gen. C. N. G.).

"Your book is emphatically one for the 'spot,' and in that respect alone, besides the very thorough way in which the ground is covered, it deserves all that has been said of it."—ROBERT N. ROLFE (Capt. N. H. N. G.).

www.ingramcontent.com/pod-product-compliance
Lightning Source LLC
Chambersburg PA
CBHW021350230426
43666CB00006B/476